Supercapacitors Based on Carbon or Pseudocapacitive Materials

Energy Storage – Batteries, Supercapacitors Set

coordinated by
Patrice Simon and Jean-Marie Tarascon

Volume 3

Supercapacitors Based on Carbon or Pseudocapacitive Materials

Patrice Simon
Thierry Brousse
Frédéric Favier

ISTE

WILEY

First published 2017 in Great Britain and the United States by ISTE Ltd and John Wiley & Sons, Inc.

ISTE Ltd
27-37 St George's Road
London SW19 4EU
UK

www.iste.co.uk

John Wiley & Sons, Inc.
111 River Street
Hoboken, NJ 07030
USA

www.wiley.com

Library of Congress Control Number: 2017933651

British Library Cataloguing-in-Publication Data
A CIP record for this book is available from the British Library
ISBN 978-1-84821-722-5

MIX
Paper from responsible sources
FSC® C013604

Contents

Introduction

I.1. Why supercapacitors?

Electrochemical capacitors, or supercapacitors (SCs), are electrochemical energy storage systems that deliver or absorb large peak power [SIM 08a, CON 99].

Figure I.1 shows the advantage of using these systems and their complementarity with batteries (LFP, in this case). The figure depicts the energy density delivered or stored according to charge time. For longer charge times (low charge rates), the battery stores 20 times the energy stored in supercapacitors. When the charge time decreases (faster charge rates), the energy density of batteries decreases, whereas that of supercapacitors remains almost stable; for charge times of a few seconds, a supercapacitor can store more energy than a battery.

The two curves intersect at about 10 s, which approximately defines the usage boundary of the two systems: applications requiring a supply of energy in short bursts (power), typically less than 10 s, should be addressed by supercapacitors, whereas batteries are better suited to applications requiring a longer energy supply >10 s.

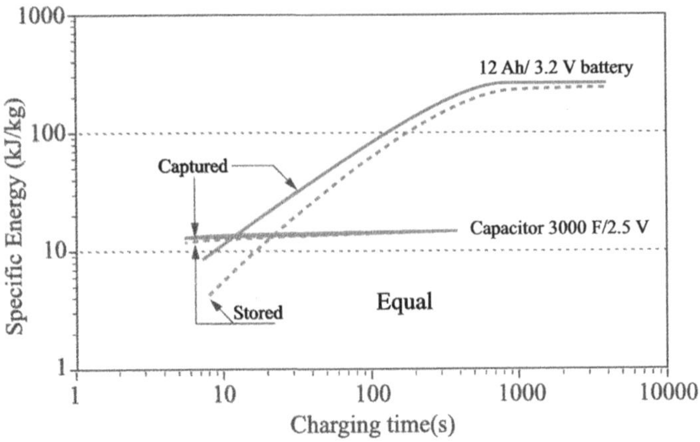

Figure I.1. *Discharge curves of a 12 Ah/3.2 V LFP/graphite Li ion battery and a 3,000 F/2.5 V supercapacitor. The battery and the supercapacitor have an identical volume [MIL 08]. For a color version of the figure, see www.iste.co.uk/brousse/supercapacitors.zip*

The specific features of supercapacitors are due to the charge storage mechanism which occurs at the surface of active electrode materials, unlike batteries that store the charge in the bulk of active materials through redox reactions. In supercapacitors, two main types of active materials can be distinguished, as they store charge in two different ways:

– the vast majority of supercapacitors currently uses porous carbon with a large specific surface area (SSA) as the electrode material; these supercapacitors are called electrochemical double-layer capacitors (EDLCs). They store charge by accumulating ions from the electrolyte at the surface of carbon electrodes under polarization [SIM 08a]; there are therefore no redox reactions involved;

– pseudocapacitive materials are the second family of materials used in supercapacitors. As their name suggests, the electrochemical signature of these materials seems to be capacitive, in the way that it is similar to that of a carbonaceous material (see Figure I.3); however, the storage mechanism is different, since the energy is stored by means of fast and reversible redox reactions, usually occuring at the (sub)surface of the material. We will come back to the storage mechanisms and the properties of pseudocapacitive materials in Chapter 4 [CON 99].

I.2. Some figures

A growing number of scientific articles containing the words "supercapacitor" or "pseudocapacitor" has been published since 2003–2004. Approximately 3,000 articles contained one of these words in 2016; this highlights the current enthusiasm for these systems. When it comes to global research, France is in fifth place, primarily due to groups in Montpellier (ICG), Nantes (IMN), Orléans (CEMHTI) and Toulouse (CIRIMAT), and all members of the French Network on Electrochemical Energy Storage (RS2E). Laboratories such as IS2M in Mulhouse, LCMCP of Chimie Paris Tech or ICMCB in Bordeaux are also strongly involved in the topic as part of RS2E.

I.3. Applications

There are two main types of applications for supercapacitors depending on the format (actual capacitance) of the cells used. In small formats (capacitance lower than 50 F), supercapacitors have been used for over 20 years in the field of power electronics, for example in the development of power buffers or to supply energy to sensors. They are also found in small tools or even in some toys. The real turning point came with their use as the power supply for emergency systems in the opening of the doors of the Airbus A380, a program that began in 2005 and was developed from the design of the aircraft. Even if it is a niche market, it has, on the one hand, demonstrated the advantage of using supercapacitors for power applications, but also the maturity, safety and reliability of this technology. Since then, the supercapacitor market has continued to develop both in small formats as well as in those with greater capacitances (several thousand Farads). Today there are many applications, most of the time in the form of modules of several tens or hundreds of volts. On tramway or trolleybus lines, they are used in network regulation, to boost charging with brake energy recovery (or the recovery of potential energy in the case of harbor cranes). Applications using supercapacitors for electric traction also exist in electric buses in the city of Shanghai [CHI 13] or even the recent electric tramway lines (2014) of the Bolloré group (*Blue Solutions company*) [OBJ 17] and the electric boat from the same Blue Solutions company [LEJ 13]. In these applications, supercapacitors are used for their capacity to recharge rapidly (within a few tens of seconds). Their autonomy is however limited by the space available to embark the SCs.

Due to significant interest in these systems, supercapacitors have also emerged in the automobile transport market. In this application, it is Citroën who was the driving force, since 2011, with the C3 and C4 including a 1.2 kW starter-alternator supplied by Continental using Maxwell supercapacitors. The starter-alternator helps to perform *stop / start* functions and regenerative braking.

Other manufacturers closely watched this technology and in 2012 Mazda embarked upon the *stop and start* venture in partnership with Nippon Chemicon.

A great deal of detailed information on the applications of supercapacitors is available at www.supercondensateur.com. This site is hosted by an enlightened amateur, passionate for new technologies and it is quite comprehensive, even if some information must be read with hindsight.

In summary, supercapacitor technology has now matured and many manufacturers explore the world of supercapacitor cells or modules. Table I.1 lists the five main manufacturers of supercapacitors in the world. The word leader in terms of production volume and range of products is Maxwell, NessCap and Nippon Chemi-con closely follow. Nippon Chemi-con devices suffer from the use of propylene carbonate (PC) as the solvent of the electrolyte, thus reducing the power density by a factor of 2–3.

CapXX specializes in high power cells (70 kW/kg) but with lower capacities (several hundred mF), primarily in flexible prismatic formats. Blue Solutions specializes in large capacitance cells and only targets applications requiring large voltages and/or capacitances (tramways, boats, buses, etc.).

Company	Country	Product	Design	Unit voltage (V) / solvent	Capacitance (F)	Module
MAXWELL	USA	Bootscap	Prismatic and cylindrical	2.7 / AN, PC	5 - 3000	>125V
NESSCAP	Korea	Nesscap	Prismatic and cylindrical	2.7 / AN, PC	3 - 5000	Up to 340V
Nippon Chemi-con	Japan	DLCap	Prismatic and cylindrical	2.7 / PC	300-3000	>200V
BATSCAP	France	Blue solutions	Cylindrical	2.7 / AN, PC	600 - 9000	>300V
CAP-XX	Australia	CAP-XX	Prismatic (2D)	2.5 AN	0.075 - 2	N/A

Table I.1. *Main supercapacitor manufacturers. Maxwell recently announced the development of a supercapacitor with a nominal voltage of 3 V*

I.4. The key parameters of supercapacitors

I.4.1. *Cell voltage V*

The maximum potential difference at the supercapacitor terminals (V_{max}) depends on the nature of the electrolyte and the current collectors chosen. Aqueous electrolytes have a (thermodynamic) theoretical stability domain of 1.2 V. They give rise to systems whose nominal voltage does not exceed 1.2 V as a general rule due to the redox reactions of the water appearing at higher voltages (except in the presence of kinetic overpotentials on the electrodes). However, the conductivity of these electrolytes is higher than those of organic electrolytes (>100 mS/cm for highly concentrated acid or alkaline electrolytes), which causes lower specific resistances for systems in an aqueous medium (≈0.5 $\Omega{\cdot}cm^{-2}$ in the best of cases). On the other hand, organic electrolytes have a much greater stability domain, approximately 3 V, but a much lower conductivity (several dozen mS/cm in the best of cases, see Chapter 2, on electrolytes). The maximum potential difference at the supercapacitor terminals can also be limited by corrosion of the current collector and/or the oxidation of the active material of the positive electrode. A typical supercapacitor electrolyte contains a solvent (either acetonitrile or PC) and a salt, usually a fluoride one (e.g. tetraethylammonium tetrafluoroborate). The operating voltages of supercapacitors with an organic electrolyte are in the range 2.7–3 V.

I.4.2. *Capacitance*

The capacitance (unit Farad, F) depends on the nature of the active materials and the electrolyte used. In aqueous media, the best carbon can reach 300 F/g, whereas in organic media, the highest capacitances are approximately 150–200 F/g. The differences observed are mainly due to the porous carbon structures and differences in the dielectric constant of the solvents. As the energy stored in the capacitor depends in particular on the capacitance value, the choice of the activated carbon is important to make a carbon/carbon supercapacitor. Pseudocapacitive materials have capacitances that can reach several hundred F/g due to surface redox reactions. In the case of these materials, the density of the prepared electrodes (mg of active material/cm^2 of electrodes) is critical, since very thin films (<1 μm) or very low electrode densities (<1 mg/cm^2) can generate very high capacitance values (F/g) but are not of any practical use when assembling cells, except in microsystems because of their low areal capacitance (hundreds of mF per cm^2). This point will be discussed later on.

I.4.3. *Series resistance R*

The series resistance R is the resistance of the supercapacitor. It mainly consists of the electrolyte resistance (approximately 60%) but also various resistances occurring from the electrodes such as the contact resistance and the intrinsic resistance of the materials. The resistance of systems operating in aqueous electrolytes is, generally, lower than that of systems in organic electrolytes due to the greater conductivity of the aqueous electrolytes than when using various non-aqueous solvents. The resistance determines the maximum power delivered by the supercapacitor as shown in equation [I.2].

I.4.4. *Energy*

The energy stored in the capacitor is given by the following equation:

$$W=\frac{1}{2}CV^2 \qquad [I.1]$$

where W is the energy (J), C is the capacitance (F) and V is the voltage (V). Systems operating in organic electrolytes, due to larger potential stability window (up to 2.7–3 V), store more energy than in aqueous systems, but with lower maximum capacitances than the latter.

I.4.5. *Power*

The power is the product of the voltage times, the current, or the energy delivered per unit of time. The parameter frequently used to characterize supercapacitors is the maximum power P_{max}. This power is a theoretical value that corresponds to the power that would be delivered during a discharge at nominal time. It is given by the following equation:

$$P_{max} = \frac{V_{max}^2}{4R} \qquad [I.2]$$

where P_{max} is the maximum power in W.

I.5. Carbon-based supercapacitors (EDLCs)

Over 90% of commercial supercapacitors use carbon as the active electrode material. They are called "electrochemical double-layer capacitors" (EDLCs).

I.5.1. *Charge storage in EDLCs*

Carbon-based supercapacitors or EDLCs store energy by accumulating ions from the electrolyte at the surface of polarized electrodes, thus charging the electrochemical double-layer capacitance [SIM 08a, CON 99], as shown in Figure I.2. The capacitance of the electrochemical double layer can therefore be described by the following equation:

$$C = \varepsilon_0 \varepsilon_r S / d \qquad [I.3]$$

where ε_r is the dielectric constant of the electrolyte, ε_0 is the vacuum permittivity (F/m), S is the geometric surface of the electrode accessible to ions (m^2) and d is the approaching distance of the ion at the surface of the carbon electrode (m). The capacitance of the electrochemical double layer is approximately 10–20 $\mu F/cm^2$ of electrode surface. The strategy will therefore be to use a carbonaceous material with high SSA, typically up to several thousand m^2/g.

When two carbon electrodes are immersed face to face into an electrolyte, the applied potential difference ΔV leads to the accumulation of a charge Q at the electrodes according to the following equation:

$$Q = C \cdot \Delta V \qquad [I.4]$$

where Q is the charge (C), C is the capacitance (F) and is a constant all over ΔV which is the applied potential difference (V).

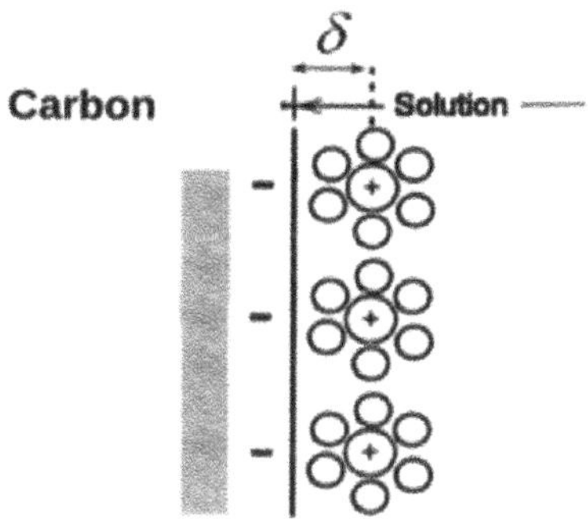

Figure I.2. *The charge of the double electrochemical layer during the negative polarization of a carbon electrode*

The adsorption of ions is fully reversible, which confers a virtually unlimited cyclability to the system. The electrodes behave as ideally polarized blocking electrodes, i.e. there are no Faradic reactions at the electrodes as the voltage increases. The maximum voltage is limited by the reduction and oxidation potential of the electrolyte and/or by the corrosion of the positive current collector (aluminum foils in organic electrolytes), or reduction or oxidation of electrode materials. The electrode/electrolyte system acts like a plane capacitor, in which the solvent plays the role of the dielectric. With the polarization being localized at the electrode/electrolyte interface, the electric field reaches quite a significant value (approximately 10^{10} V/m), with the thickness of the electrochemical double layer being only in the range of a few nanometers. This confinement to the nanometric scale of the electrostatic energy is the reason behind the very high density of charge stored. Energy densities of about 100 kJ/L can be obtained, a value much greater than that of classical dielectric capacitors, but much lower than in batteries. An EDLC supercapacitor consists of two capacitors in series: a positive electrode associated with anions in solution and a negative electrode associated with cations in solution. Table I.2 lists the characteristics of batteries and supercapacitors.

	Supercapacitors	Batteries
Energy density (Wh/kg)	1–5	20–200
Power density (kW/kg)	2–15	<2
Charge time (s)	1–30	1–5 h
Discharge time (s)	1–30	0.2–10 h
Cyclability (cycles)	10^5–10^6	300–10,000
Lifetime (years)	>30	5

Table I.2. *Comparison of the performance of batteries and supercapacitors [MIL 08]*

The differences observed between the two kinds of devices lie in the different charge storage mechanism: on the active material surface in the case of supercapacitors, and in the bulk of the active materials for batteries. This surface storage also explains the differences in cyclability since there are only very low mechanical stresses associated with charge or discharge phenomena in ELDC supercapacitors.

I.5.2. *Electrochemical characteristics of EDLCs*

The electrochemical characteristics of a carbon/carbon supercapacitor are shown in Figure I.3. They can be explained using the basic equation:

$$Q = C \cdot \Delta V, \text{ with } Q = I \cdot t \qquad [I.5]$$

where Q is the charge (C), C is the capacitance (F), ΔV is the potential difference (V), I is the current (A) and t is the time (s).

During galvanostatic cycling (charge and discharge cycles at constant current I, Figure I.3(a)), the change in voltage versus time is therefore linear. The slope is inversely proportional to the capacitance since $C = I/(dV/dt)$ and the gravimetric capacitance can be determined per gram of electrode material.

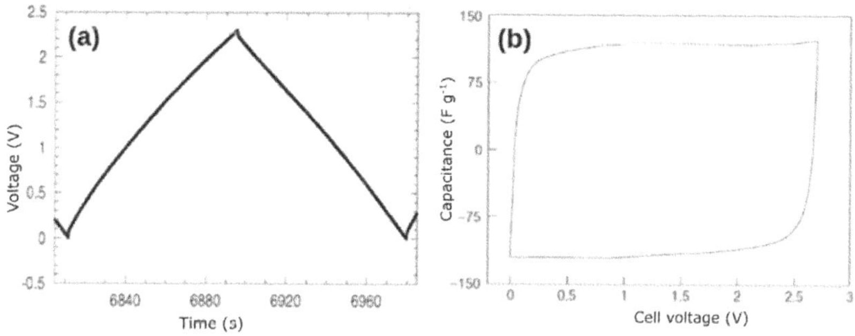

Figure I.3. *a) Constant current galvanostatic charge/discharge plot; b) cyclic voltammotry of a supercapacitor [SIM 08a]*

During cyclic voltammetry experiments, a constant potential scan rate (dV/dt = cste) is used; therefore, the theoretical voltammogram of a supercapacitor is rectangle shape since C and dV/dt are constant. The experimental voltammogram (Figure I.3(b)) is consistent with the theory. The capacitance of the system is calculated from equation I.5 by integrating the area under the CV that corresponds to the charge Q, and dividing Q by the width of the potential window.

Another type of electrochemical characterization is the electrochemical impedance spectroscopy. This technique gives changes in capacitance and resistance of supercapacitors depending on the frequency, i.e. according to the kinetic of use (e.g. pulse use). When supercapacitors use porous electrodes such as porous carbons in EDLCs, the capacitance depends on the frequency: it shows its maximum at low frequency (DC conditions), and decreases with faster discharge rates (higher frequencies). Figure I.4(a) shows the impedance plot in the Nyquist plot (Z″ vs. Z′) of a 2.5 F/2.5 V laboratory supercapacitor cell device. Figure I.4(b) shows the change in capacitance versus frequency calculated according to a Cole–Cole model [TAB 03].

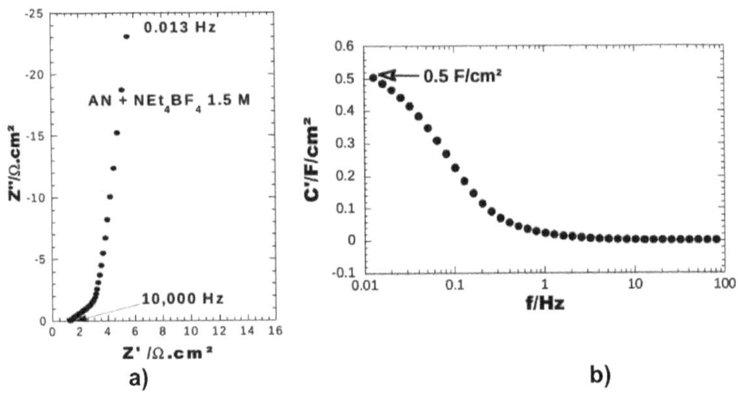

Figure I.4. *a) Nyquist plot of a 2.5 F/2.3 V supercapacitor between 10 kHz and 13 mHz at 2 V [TAB 03]; b) variation in capacitance versus the frequency obtained from a Cole–Cole model [TAB 03]*

The nature of the components used for the experiment (electrolytes, carbon, etc.) as well as the active material film thickness significantly affect the change in capacitance with the frequency (pulsed applications).

I.6. Challenges for supercapacitors

Today, supercapacitors exhibit power densities compatible with most of the applications. The greatest challenge for these systems is to increase their energy density from 5 to 6 Wh/kg (today's values) to beyond 10 Wh/kg. The energy density being given by $W = \frac{1}{2} \cdot C \cdot V^2$ as mentioned previously, it can be improved by either increasing the capacitance or enlarging the voltage. There are various strategies to increase the capacitance and/or the operating voltage.

I.6.1. *Increasing the capacitance of EDLCs*

The capacitance in EDLCs primarily depends on the carbon/electrolyte interface. It can be optimized in two different ways:

– adapting the carbon pore size distribution to the size of the ions from the electrolyte, or more generally controlling the pore size distribution of the carbon;

– functionalizing the carbon by electrochemically active surface groups that will contribute to the total capacitance.

These strategies will be further discussed in Chapter 1.

I.6.2. *Increasing the cell voltage*

Tailoring electrolytes with high ionic conductivity and the widest electrochemical stability window possible is clearly a major objective for the electrochemical energy storage community; However, there are numerous limitations and limited methods to reach these goals. Carbon-based supercapacitors do not use redox reactions to store charges and as they do not need to form a solid electrolyte interphase to operate in the same way as lithium-ion batteries, they offer an additional degree of freedom with regard to batteries to develop high-voltage electrolytes, particularly when using ionic liquid mixtures. Chapter 2 presents the main results obtained with today's electrolytes, and depicts the perspectives to develop new electrolytes for SCs.

I.6.3. *Increasing the capacitance: pseudocapacitive materials*

Pseudocapacitive materials store charges through rapid and reversible redox reactions occuring at the (sub)surface of the electrode materials. In this way, redox reactions are not limited by diffusion and have a similar electrochemical signature to that of EDLCs (see Chapter 3). Using surface redox reactions to increase the capacitance of materials is interesting since the Faradaic capacity arising from redox reactions will always be greater than the double-layer capacitance arising from ion adsorption. One way to increase the capacitance of materials is therefore to use nanostructured pseudocapacitive materials. This will be discussed further in Chapter 4.

I.6.4. *Hybrid devices*

Hybrid devices couple a battery type electrode and a supercapacitor type electrode. The battery electrode stores charge in its bulk by redox reactions and subsequently increases the energy density of the hybrid device. The associated supercapacitor electrode improves the power of the system. The most frequently studied combinations use nowadays in hybrid devices operated either in aqueous or organic electrolytes are C/NiOOH, C/PbO_2, C/MnO_2, or lithiated graphite/porous carbon, respectively. The latter device (lithium ion capacitor) uses a Li ion battery anode that also helps to increase the operating voltage beyond 3 V. This will be discussed further in Chapter 5.

1

Electrochemical Double-Layer Capacitors (EDLC)

Carbon is the material of choice for supercapacitor electrodes (electrochemical double-layer capacitors [EDLCs]), because it meets three important requirements needed for active materials for electrochemical energy storage applications, which are:

– electrical conductivity;

– electrochemical stability;

– high surface area, that can be easily developed by physical or chemical methods.

The electrical conductivity of carbon, even when nanostructured, can reach several S/cm, which allows electrochemical polarization without ohmic drop limitations. Carbon is also stable over a wide potential range in aqueous or non-aqueous electrolytes, meaning that the voltage is only limited by redox reactions of electrolytes during the polarization of the electrodes. Finally, there are different forms and structures of carbon whose specific surface areas (m^2/g) can be easily controlled.

1.1. The different forms of carbon

Figure 1.1 shows the different forms of carbon used in supercapacitors [SIM 13].

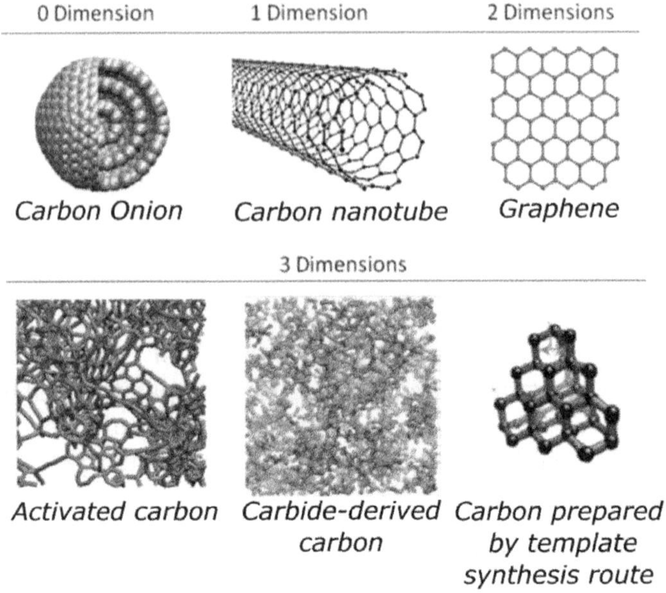

Figure 1.1. *Different carbon forms used in supercapacitors, ranging from 1 to 3 dimensions [SIM 13]. For a color version of the figure, see www.iste.co.uk/brousse/supercapacitors.zip*

Carbon structures for EDLC applications must develop a large surface area (in m^2/g) in order to maximize the capacitance (number of ions adsorbed). Equation [I.3] suggests that the capacitance changes linearly with the specific surface area, which explains why, for years, the carbon specific surface areas was the key parameter to control. Thus, the vast majority of carbon structures proposed as electrode materials are activated carbon structures.

Activated carbon with a large surface area can be prepared from natural precursors such as wood, coal or plants (coconut, trees, fruit cores, peat, etc.) [PAN 06]. Other synthetic precursors such as phenolic resins can also be used, but for cost reasons natural precursors, available in large amounts, are mainly used. Activated carbon can be prepared by physical or chemical processes. Typically, a physical process starts with the carbonization of the precursor at low temperatures (e.g. 400 °C) under inert atmosphere; it is further activated by heat treatment at 1,000 °C under oxidizing atmosphere (CO_2, water vapor, etc.). During chemical processes, both the carbonization and activation are achieved simultaneously at ~700 °C in a controlled atmosphere in the presence of a chemical agent ($ZnCl_2$, phosphoric acid, etc.) helping to develop the porosity by catalyzing the activation, dehydration, carbonization and restructuring.

None of these activation processes can accurately control the sizes of the pores formed. In general, such processes only give access to an average pore size, which is the parameter to control. The activation step develops the porosity inside the carbon grains, which ultimately increases the total carbon surface area, thus allowing it to store more charges. The specific surface area is measured using a gas sorption technique, which is based on the reversible adsorption of dinitrogen gas or argon in the pores of the studied materials. Using these measurements, it is possible to determine the average pore diameter as well as the pore size distribution. Three porosity domains are defined according to the pore diameter (Figure 1.2):

– micropores: pore diameter < 2 nm;

– mesopores: 2 nm ≤ pore diameter < 50 nm;

– macropores: pore diameter ≥ 50 nm.

Figure 1.2. *IUIPAC classification of different pore sizes*

It is important to mention that adsorption of CO_2 is much more adapted to the study of microporous carbon.

Since 2006 [CHM 06], activation processes have been the main way to develop the surface of carbonaceous materials (pore size and pore size distribution), which plays an important role in the capacitance values. Other synthesis techniques are also described in the literature, aiming to better control the pore size of carbonaceous materials. As an example, organized mesoporous carbons can be prepared via the *template* method [MA 13, NIS 12]. This method uses a rigid siliceous mesoporous matrix impregnated with a carbon-containing precursor (polymer or copolymer). A *hard template* means that the host template must be dissolved to obtain the final product – the carbon. It is then calcined and the silicon skeleton is dissolved to obtain the inverse replica of the matrix, which is a mesoporous carbon [VIX 04, GAD 08].

Another method involves the use of structuring agents to prepare mesoporous carbons, such as surfactant block copolymers (Pluronic© type) [MA 13]. The block copolymer is mixed with a carbon precursor (e.g. a phenolic resin) containing –OH functional groups. The resin forms hydrogen bonds with the hydrophilic part of the block copolymer in an alcoholic environment. At the same time, micelles self-organize around hydrophobic parts and form spheres or cylinders depending on the surfactant used. A mild heat treatment causes the resin to polymerize around the micelles and the subsequent carbonization destroys the surfactant and transforms the precursor into carbon [MEN 05]. This *soft template* method accurately controls the pore size; however, this technique and the silica *templates* used are costly.

New carbon families have appeared in the field of supercapacitors, such as carbide-derived carbons (CDCs) obtained by chlorination of metallic carbides such as TiC [CHM 06] or graphene and activated graphene [ZHU 11]. Onion-like carbons (OLCs) are equally interesting. OLCs show modest surface area (500 m^2/g), but the key feature is that this surface is completely accessible to electrolyte ions since these nanoparticles (5 nm diameter) are non-porous [SIM 13]. OLCs are obtained by the graphitization of nanodiamonds (heat treatment at 1,500 °C). The achievable capacities remain low, as the surface is not highly developed, but these materials can deliver large power due to the full access to the carbon surface.

Carbon nanotubes offer similar power characteristics to those of OLCs as only the external surface of the tubes is available for ion adsorption.

1.2. Increasing the capacitance of microporous carbon

Some results in the literature had suggested the importance of micropores in the storage of charges in porous carbon in early 2000 [RAY 06a], but it was only in 2006 that their role was clearly described with the use of CDC [CHM 06] by a RS2E group in collaboration with Drexel University (USA). CDCs are carbons with sub-nanometer pores (< 1 nm), obtained via the chlorination of metallic carbides at temperatures between 400 and 1,000 °C according to the reaction below:

$$MC + nCl_2 \rightarrow MCl_{2n} + C_s \quad [1.1]$$

The advantage of this method is the ability to prepare carbons with fine-tuned, controlled pore sizes and pore size distributions, since the porosity is developed through the elimination of the metal M atoms. Porous carbon prepared from TiC has porosities between 0.6 and 1.2 nm. CDCs have been used as *model materials* and tested in an electrolyte (NEt_4BF_4 1 M in acetonitrile), with the size of solvated ions being 1.7 and 1.5 nm for NEt_4^+ and BF_4^-, respectively. The results in Figure 1.3 show that the micropores, with sizes smaller than the diameter of solvated ions, play an important role in storing the charge. From a fundamental point of view, it means that ions must be partially desolvated to access these pores.

Moreover, despite the confinement of ions in these narrow pores and contrary to conventional beliefs, the ion transport and dynamics in these nanopores are only slightly affected: the capacitance decreases by only 10% when the current density increases from 5 to 100 mA/cm² [CHM 06].

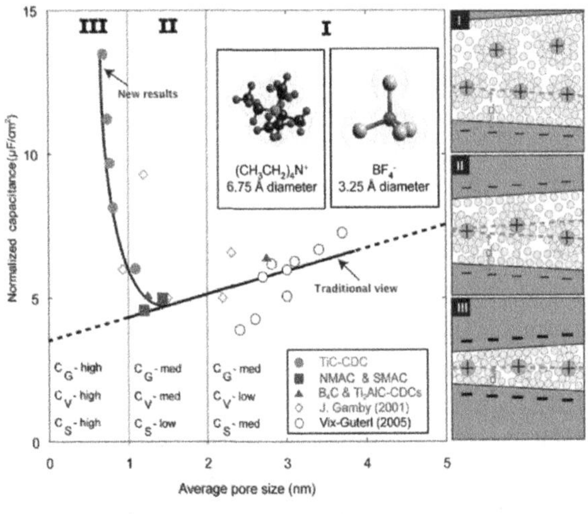

Figure 1.3. *Variation in normalized capacitance according to pore size of different carbons; nanoporous CDCs achieved high capacitance values in subnanometer pore sizes [CHM 06]. For a color version of the figure, see www.iste.co.uk/brousse/supercapacitors.zip*

Another major finding was obtained by studying the adsorption of ions in CDCs using a solvent-free pure ionic liquid electrolyte, 1-ethyl-3-methylimidazolium bis(trifluoro methylsulfonyl)imide (EMI^+, $TFSI^-$) [LAR 08]. In this electrolyte, the maximum charge storage capacity was obtained when the carbon pore size was close to the size of the ions, meaning that the ions can access pores of the same size. These results significantly contradict the way the storage of charges (adsorption of ions) has been traditionally described in porous carbonaceous materials, with ions being adsorbed to each side of the pore walls.

All these results attracted the attention of theoreticians who rapidly proposed theories to explain the origin of this increase in capacitance in subnanometric pores [KON 11a, KON 11b]. The simulations confirmed the increase in capacitance in nanometer and sub-nanometer pores, but could not quantitatively reproduce the magnitude of this increase. This difference was primarily due to the diversity of ideal carbon structures used (carbon graphite, carbon nanotubes) and to the different methods used (mathematical approaches, Monte Carlo, reverse Monte Carlo, molecular dynamics).

It is in this context that two teams from the RS2E [MER 12] began to study the adsorption of ions in carbon nanopores by coupling the simulation by molecular dynamics and experiments, standing out from the literature:

– by using real carbon structures (CDCs);

– by adding an electrolyte reservoir between the two electrodes;

– by applying a constant voltage and not a constant charge.

By using an ionic liquid as the electrolyte (BMI, PF_6) and a CDC with a pore size of 0.9 nm, these studies showed that ions accessed pores with sizes similar to the ions in the absence of any polarization (Figure 1.4). Under polarization, the variation in the total number of ions present in the pores ($N^+ + N^-$) revealed that there was ionic exchange between the electrolyte bulk and the electrolyte in the pores. As a result, the mechanism for the increase in capacitance is therefore not based on a simple forced electrosorption phenomenon. The number of ion coordinations changes from seven in the electrolyte bulk to four in the pores, confirming a change in environment within the pores (desolvation). In the presence of a solvent (acetonitrile), the simulations show that ions replace the acetonitrile molecules even in the most confined pores, while in the pure ionic liquid the ions do not have access to these confined pores [MER 13]. This effect is amplified when a potential difference is applied.

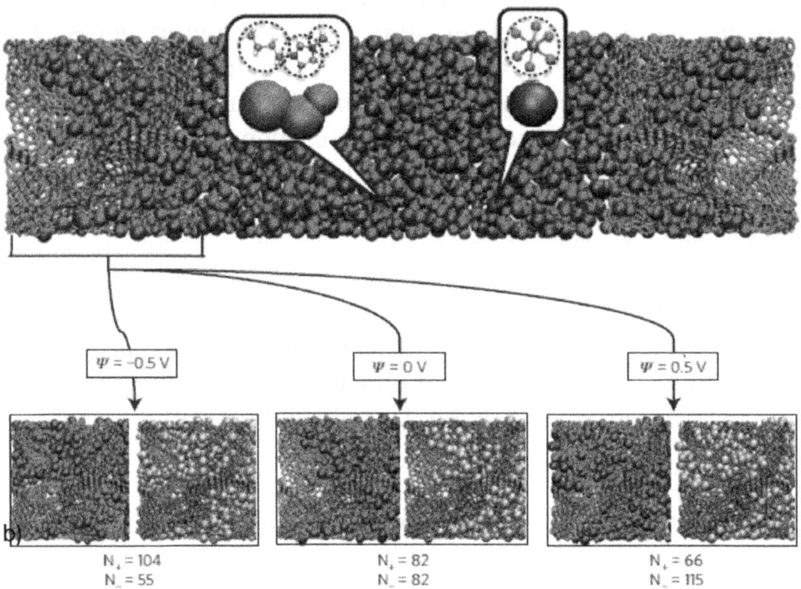

Figure 1.4. *a) Simulation of a cell with two CDC electrodes (0.9 nm pore size) in a BMI-PF6 electrolyte. The cations are in red and the anions in green; b) Electrode structure under different applied voltages; for each voltage, the ions adsorbed in the pores (left) and the charge of the carbon atoms (right) [MER 12]. For a color version of the figure, see www.iste.co.uk/brousse/supercapacitors.zip*

These results led to a better understanding of the mechanisms of ion transport and adsorption in confined pores. Today, this has positioned the RS2E at the forefront of research in this field. The work currently carried out focuses on modeling the charge dynamics of electrodes. The aim is to model electrodes and devices using the modeling of the charge and discharge of electrodes cut into slices of only a few nanometers in thickness. Initial results seem promising [PEA 14] and show that this approach is pertinent.

1.3. Activated microporous carbon

Since microporous carbon have nanometer and sub-nanometer pore sizes, they are difficult to synthesize using template methods; combined with cost issues, this explains why the vast majority of carbons currently used in supercapacitors are activated carbons specifically prepared with a *mean pore size* of approximately 1 nm. The companies Kuraray, Kansaï Coke and Japan Envirochemical are currently the global leaders in the fabrication of microporous carbons for supercapacitors. To a lesser extent, Osaka Gas Chemical (Japan), Meadwestvaco, Kabot (USA), Norit and Timcal (Switzerland) can also be mentioned. Controlled activation for the preparation of microporous carbons has considerably increased the capacitance of the activated carbons; commercial microporous carbons such as YP50F (Kuraray) currently show capacitances of 130–140 F/g in organic electrolytes. In the literature, capacitances greater than 200 F/g are now being demonstrated [WEI 12a]. The capacitances of carbon has doubled over the last 10 years following work carried out on microporous carbons, which is quite remarkable and highlights the rapid progress in the development of SCs.

The RS2E has not been left behind in this field with, among others, work on synthesis methods of carbon activated using various precursors including biomass [MYS 12, RAY 13].

1.4. Hierarchical porous carbon

Highlighting the importance of micropores in the mechanism of capacitive storage has renewed interest in the synthesis of carbons with controlled pores and pore sizes. In particular, the concept of hierarchical porosity has emerged in this field of application. The concept is based on the preparation of carbons that do not have a uniform porosity, but a bimodal pore size distribution, for example micro-/mesoporous. In these structures, the micropores provide the capacitance and the mesopores, somewhat larger, act as an electrolyte reservoir to facilitate ion transport.

The vast majority of synthesis methods detailed in the literature to prepare carbons with hierarchical porosities use the template method. The advantage here is the ability to obtain materials with highly controlled pore sizes, but the disadvantage is that the structures are meso-/macroporous due to the challenge when preparing microporous carbons (pores of approximately 1 nm) using this method. The electrochemical performances of these structures are still, however, quite modest with capacities of approximately 100 F/g [KUB 13].

An original method to synthesize micro-/mesoporous hierarchical porous carbons was recently proposed by Gao *et al.* [GAO 13b, TSA 13a]. Figure 1.5 shows the approach used to synthesize porous silicon carbide (SiC).

A carbon precursor (resol, which is a mixture of phenol + formaldehyde) is added to an alcoholic solution containing a mixture of block copolymers and a SiO_2 precursor. Carbonization at 800 °C under nitrogen leads to the preparation of a SiO_2/carbon composite. The SiO_2/carbon composite is transformed into SiC by reduction in the presence of Mg powder.

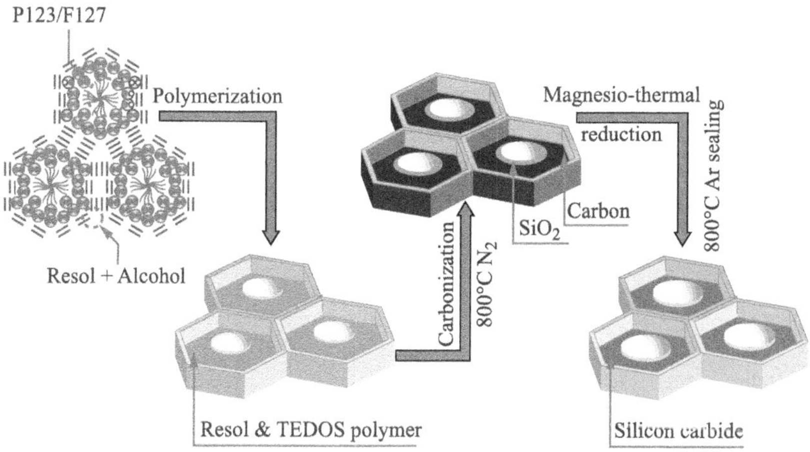

Figure 1.5. *Preparation of mesoporous silicon carbide SiC [TSA 13a]. For a color version of the figure, see www.iste.co.uk/brousse/supercapacitors.zip*

The resulting compound is therefore a mesoporous carbide, SiC, whose pore size can be adjusted according to the SiO_2/C ratio.

The final stage is to generate the microporosity through the chlorination of the carbide according to:

$$SiC_{\text{(solid)}} + 2\ Cl_{2\,\text{(gas)}} \rightarrow SiCl_{4\,\text{(gas)}} + C_{\text{(solid)}} \quad [1.2]$$

The resulting carbon shows a high specific surface area (1,600 m^2/g), and a bimodal pore size distribution with peaks at 1 and 4 nm [TSA 13a]. The electrochemical performance shows a modest capacitance of 80 F/g, due to the presence of mesopores, but with remarkable power performances, with a capacitance decrease at 3 V/s limited to 30% of the capacitance at low scan rate of 0.01 V/s (see Figure 1.6). This validates the proposed approach since controlling the pore size distribution allows the power of the material to be adjusted.

Despite this, the synthesis method is still difficult to scale-up (due to the cost and complexity of synthesis), it has the unique advantage of controlling the pore size at the microporous or mesoporous level. A wide range of carbon structures is therefore achievable, and it now becomes possible to propose carbon structures depending on the electrolyte used (and mainly the viscosity, e.g. ionic liquids). This point will be discussed in Chapter 2 (electrolytes).

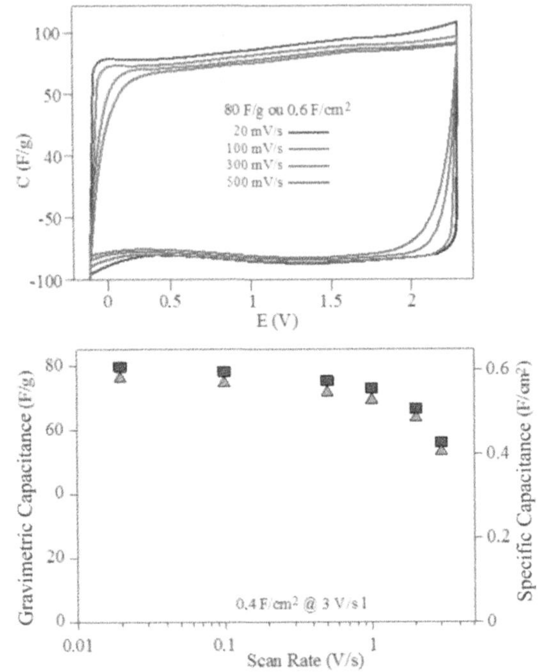

Figure 1.6. *Cyclic voltammetries and power curves of carbon electrodes with a micro-/mesohierarchal porosity [TSA 13a]. For a color version of the figure, see www.iste.co.uk/brousse/supercapacitors.zip*

1.5. Graphene

Graphene meets all the key features needed to be used as supercapacitor electrode material: a high electrical conductivity (> 1 S/cm), large electrochemical stability window (at least similar to that of porous carbons) and very high specific surface area (2,800 m^2/g). The two main issues that practically limit the performance of graphene are:

– the recombination of graphene layers;

– the low density of graphene.

Recombination is the stacking back of graphene sheets, which reduces the usable surface. The low density comes from the 2D nature of this material.

Two approaches have recently been proposed in the literature to tackle these issues. The first one was proposed by Ruoff and co-workers [ZHU 11] in 2011, who prepared porous graphene. Graphite oxide (GO) was exfoliated by microwaves and reduced to MEGO by the addition of hydrazyne. After washing in hot KOH solution, the graphene layers are oxidized and microporous graphene is obtained, with an average pore size of 1.5 nm. The performance obtained is excellent since the capacitance reaches 150 F/g in an organic medium (acetonitrile + $EMIBF_4$). However, the low density of activated graphene is the main limitation of this material, which results in modest volumic capacities (10–20 F/cm^3).

The second approach was reported in 2013 by Yang *et al.* [YAN 13]. It involves the densification of graphene to increase its volumic capacitance by preparing graphene from reduced exfoliated GO in an aqueous medium. The suspension is then filtered and a gel formed by graphene layers with water molecules trapped between the layers is obtained (hydrogel). The gel is then placed in a solution containing a volatile compound (solvent such as water or acetonitrile, propylene carbonate, etc.) and a non-volatile compound (salt such as H_2SO_4 or an ionic liquid $EMI\text{-}BF_4$). Over time, the water of the gel is exchanged with the liquid mixture. The volatile solvent is then removed by heat treatment, which also leads to the densification of the graphene film due to capillary forces. At the end, a densified graphene film containing only the non-volatile salt (H_2SO_4 or $EMIBF_4$, for instance) is obtained. Figure 1.7 shows the different stages from the hydrogel [TSA 13a].

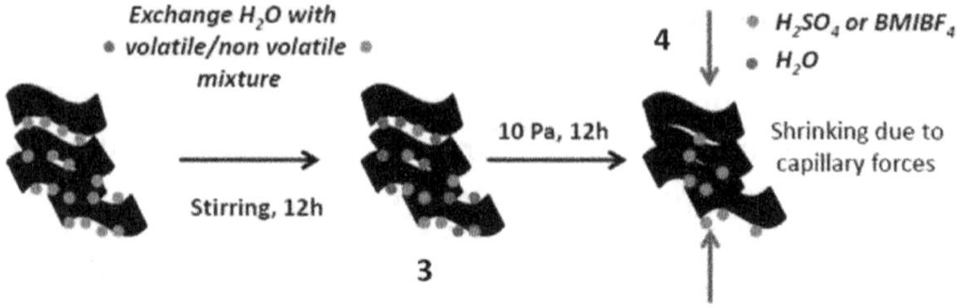

Figure 1.7. *Schematic representation of the solvent exchange in exfoliated graphene films [YAN 13]. For a color version of the figure, see www.iste.co.uk/brousse/supercapacitors.zip*

The electrochemical performance greatly improved as the volumetric capacitance reached 250 F/cm^3, with dense films (apparent densities > 1 g/cm^3, see Figure 1.8); this is an important step in the use of graphene in SCs since one of the main barriers is the low density of the electrodes prepared (low volumetric capacitance).

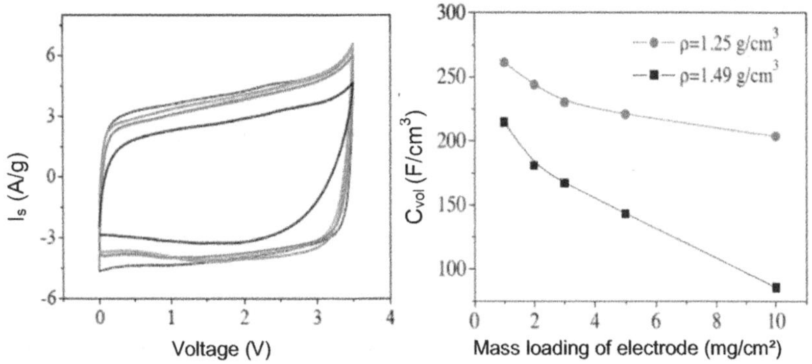

Figure 1.8. *Electrochemical performances of densified graphene films in an EMI-BF4 electrolyte in ACN. The graphene films have been previously exchanged with EMI-BF4 according to the protocol described above [YAN 13]. For a color version of the figure, see www.iste.co.uk/brousse/supercapacitors.zip*

This synthesis route allows for the preparation of dense graphene films compatible with commercial applications. The limitation will certainly still be the drying of the hydrogel films to remove traces of water (cycling stability). This approach can also be applied to the densification of powder films.

1.6. Reducing costs: carbons in aqueous media

As with batteries, one of the main challenges in the field of supercapacitors is reducing the cost. Although organic electrolytes offer the best performance in terms of energy density and cell voltage due to their extended electrochemical stability domains, the use of aqueous electrolytes can greatly reduce production costs, as well as being an environmentally friendly alternative to organic solvents. Carbon capacitance is also higher in aqueous media than in organic electrolyte because of the electrolyte characteristics (dielectric constant, viscosity, wettability, ion size, etc.). The main challenge aqueous-based carbon/carbon systems are facing is therefore increasing the energy density, that is to say, increasing the operating voltage.

In aqueous (acidic or basic) electrolytes, carbon shows significant extra capacitive contributions due to the presence of oxygen-containing functional groups, which greatly contribute to its capacitance by adding pseudocapacitance to the double-layer capacitance [HUL 05, HUL 09, RAY 06b]. The presence of other heteroatoms such as P or N also leads to pseudocapacitive contributions [HUL 05, HUL 09, RAY 06b]. The limit of this approach is undoubtedly the loss of stability in cycling due to the presence of these surface functions if the content exceeds a few percent.

Many studies have been carried out in France on this subject by the Beguin and Raymundo-Piñero group

(CEMTHI). They have shown that in neutral electrolytes, the presence of kinetic overvoltages for proton reduction and water oxidation in the confined micropores of porous carbons, greater than in a KOH or H_2SO_4 medium, allowed operating voltages above 1.5 V [DEM 10, KHO 10, GAO 12a].

Figure 1.9 shows the cyclic voltammetry of an activated carbon electrode in a Li_2SO_4 2 M electrolyte (pH 6).

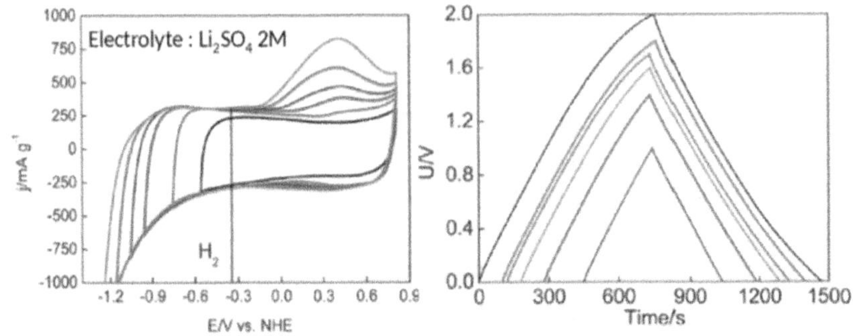

Figure 1.9. *a) Volumetric cycling of an activated carbon electrode in Li_2SO_4 2 M electrolyte. We can see the high overvoltage for H_2 evolution; b) associated galvanostatic cycling [GAO 12a]. For a color version of the figure, see www.iste.co.uk/brousse/supercapacitors.zip*

The maximum cell voltage is limited by the positive electrode. When the cell voltage is too high, carbon is oxidized to CO_2 at the active sites. By pre-oxidizing the activated carbon in oxygenated water to fully oxidize – and thus neutralize – the active sites and control the surface functions, the maximum operating voltage is increased to 1.9 V, with a 10% loss in capacitance over 10,000 cycles.

Aqueous electrolytes therefore offer interesting opportunities to supercapacitors due to the existence of overvoltages for water electrolysis (O_2 and H_2 evolution) in micropores; as a result, commercial carbon/carbon products are being finalized at Hutchinson Company. However, controlling the surface functions (passivation of active sites such as described above) to ensure performance stability over time (cyclability) is still a challenge. Another critical feature is the aluminum current collector. Unlike in organic electrolytes, aluminum (Al) anodization occurs in neutral aqueous solutions; in addition, Al is corroded in strongly acidic or alkaline medium and therefore cannot be used. Ni or Ti metals are also unsatisfactory in terms of stability or interface resistance, which explains why noble metals (Au or Pt) are used as current collectors at lab scale. Stainless steel is still not satisfactory because of its high density and cost. Other options deal with the use of Ti (with the drawback of its electrochemical stability) or the design of protective coatings on Al or Cu (corrosion is then an issue induced by the presence of scratches).

1.7. Functionalized carbon

As explained previously, the double layer capacitance of carbon is closely related to the pore size distribution, the specific surface area and the electrolyte used. There is therefore a theoretical limitation to the capacitance accessible at the level of the electrochemical double layer. To increase this capacitance, a solution is to combine the double layer capacitance of a carbon electrode with the Faradaic capacity resulting from redox reactions at the surface of the electrode. The charge is then stored via two distinct mechanisms: electrostatic and Faradaic. To preserve the key features of carbon electrodes (electrical conductivity, cyclability), it is important to not drastically modify the carbon surface. For this reason, the material providing the additional Faradaic capacity must not cover the carbon surface as a thick and insulating film, but instead decorate it to allow electroyte ions to access the pristine carbon surface.

The grafting of quinone functions has been widely described in the literature. Quinone molecules play an important role in biological systems, in chemical synthesis, or as active molecules in electrochemical systems such as supercapacitors [POG 11b, LEB 11b]. For the latter, although several methods have been studied regarding modifying surfaces with quinones, we will specifically describe grafting by the reduction of diazonium cations [POG 11a].

The reduction of a diazonium cation occurs following the spontaneous transfer of an electron from the substrate or from cathodic polarization of the substrate. This reduction reaction leads to the evolution of nitrogen dioxide gas and the formation of a radical, which further reacts with the surface atom to form a covalently attached radical [COM 09].

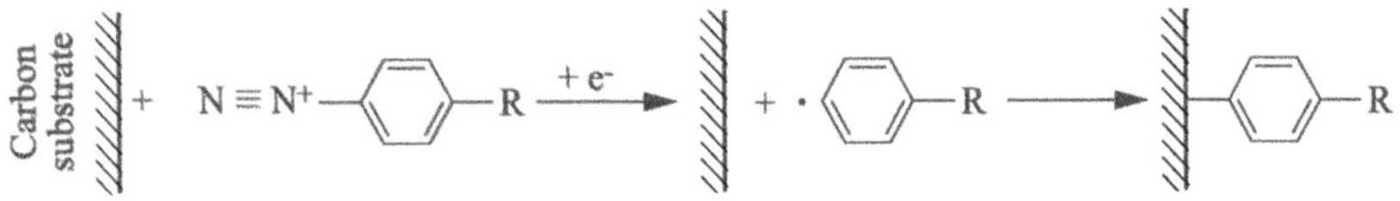

Figure 1.10. *Diagram of the grafting process using diazonium cations*

This reaction can be achieved by introducing a diazonium salt (previously prepared or purchased if available) into an acidic or organic solution in contact with the substrate to be modified [BÉL 11]. Alternatively, the diazonium cation can be generated *in situ* from the corresponding amine.

In supercapacitors, carbons functionalized by quinone molecules have demonstrated promising electrochemical properties for use as an electrode active material in acidic electrolytes [POG 12, ALG 09]. The modification of the carbons with quinone moieties leads to a significant increase in the specific capacity of the carbon due to the contribution of the redox reaction of the immobilized molecules, which generally involves one electron per unit of carbonyl radical, as illustrated in Figure 1.11(a) for anthraquinone (AQ) in acidic electrolyte.

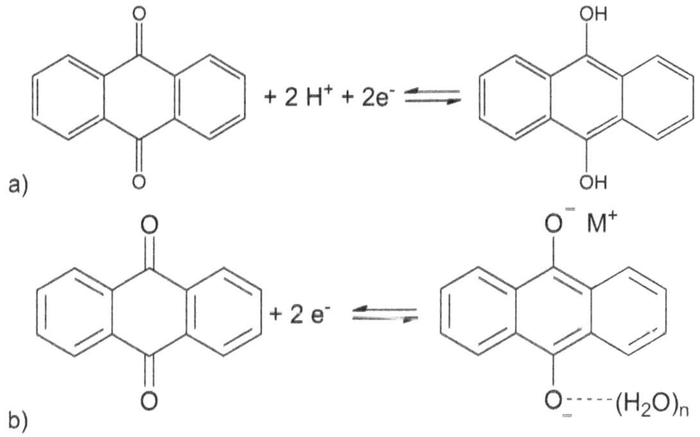

Figure 1.11. *Redox reaction of anthraquinone: a) in an acidic medium; b) in an alkaline medium where the charges of the dianionic reduced form are compensated by hydrogen bonds to the protons or alkaline cations (Na+ or K+) [LEC 14]*

The number of studies dealing with the use of AQ in alkaline electrolytes is more limited [WEI 12b, CHO 11]. The redox reaction of AQ in an alkaline electrolyte is illustrated in Figure 1.11(b). It is achieved via the transfer of two electrons and the charge compensation by cationic species (M^+) [QUA 07].

In sulfuric acid electrolyte, activated carbon electrodes functionalized with AQ molecules achieved capacitance twice than that of pristine (unmodified) carbon. Additionally, the grafting has almost no influence on the double layer capacitance of the pristine carbon electrode except at very high grafting yields, and power performances close to those of pristine carbon electrodes were obtained [POG 11b]. Note that reporting gravimetric capacitance (F/g) is unsuitable for these electrodes; alternatively, C/g (or A.s/g) is more appropriate due to the Faradaic electrochemical signature of the grafted AQ molecules. However, the cyclability is still the limiting factor of these devices because some AQ molecules are just physically adsorbed onto the carbon surface and are gradually released into the electrolyte upon cycling. Another major drawback of this approach deals with the need for washing the functionalized carbon powders, sometimes with large amounts of toxic solvents. This is why new approaches to develop more stable electrodes are being developed [LEC 13, LEC 14, LEC 15].

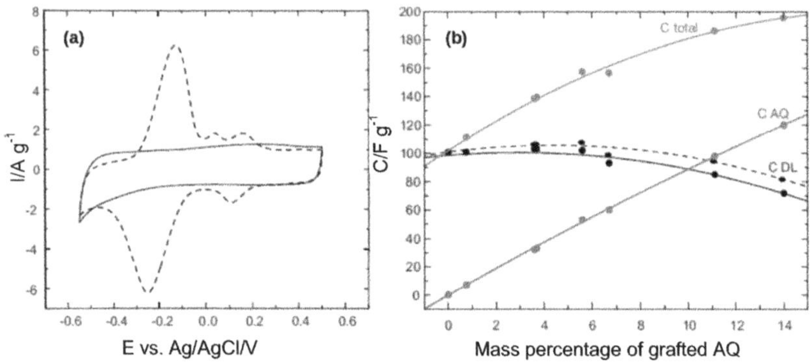

Figure 1.12. *a) Cyclic voltammogram of an activated carbon electrode (black line) and the same electrode functionalized with 11 wt. % of anthraquinone (dotted line) in 0.1 M H_2SO_4 at 10 mV/s; b) total capacitance (red), capacitance due to anthraquinone (green) and the double layer capacitance of the modified carbon electrode according to the wt. % of grafted molecules [POG 11b]. For a color version of the figure, see www.iste.co.uk/brousse/supercapacitors.zip*

Another challenge is to combine electroactive molecules with complementary electrochemical stability domains in order to assemble hybrid systems; this can be achieved by combining two different functionalized carbon electrodes. The catechol is an interesting molecule to be grafted onto the positive electrode, which will be associated with a negative electrode functionalized with *AQ* (Figure 1.13) for practical use in an aqueous electrolyte.

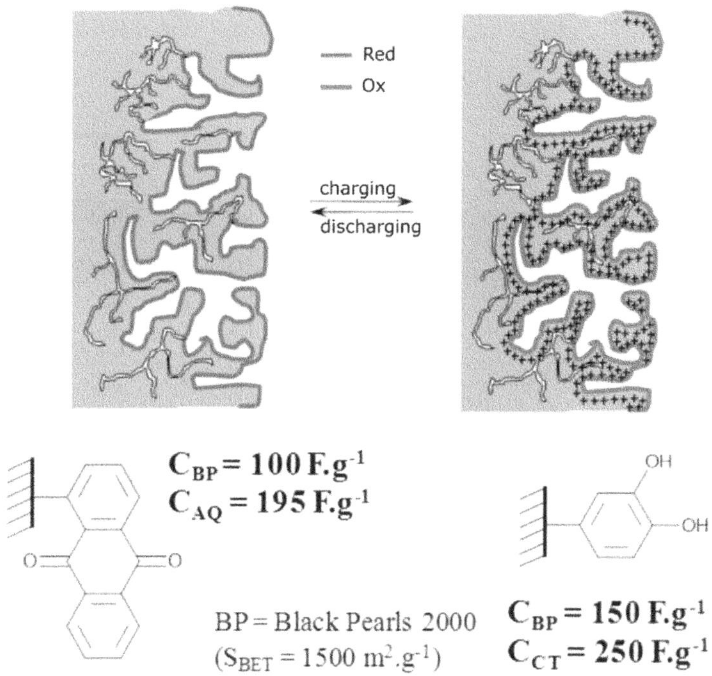

Figure 1.13. *Hybrid system with a negative carbon electrode grafted with anthraquinone (AQ) and a positive electrode grafted with catechol [POG 12]. For a color version of the figure, see www.iste.co.uk/brousse/supercapacitors.zip*

It was recently shown that it is possible to apply the same strategy in an organic medium [LEB 14] with, in this case, a significant increase in the cell voltage and the same benefit with regard to the capacitance obtained from the functionalization of the two carbon electrodes. Overvoltage for electrolyte reduction/oxidation has also been observed, leading to an increase in the electrochemical windows of the carbon electrodes; more work is needed to obtain a better understanding of the reaction mechanisms (Figure 1.14).

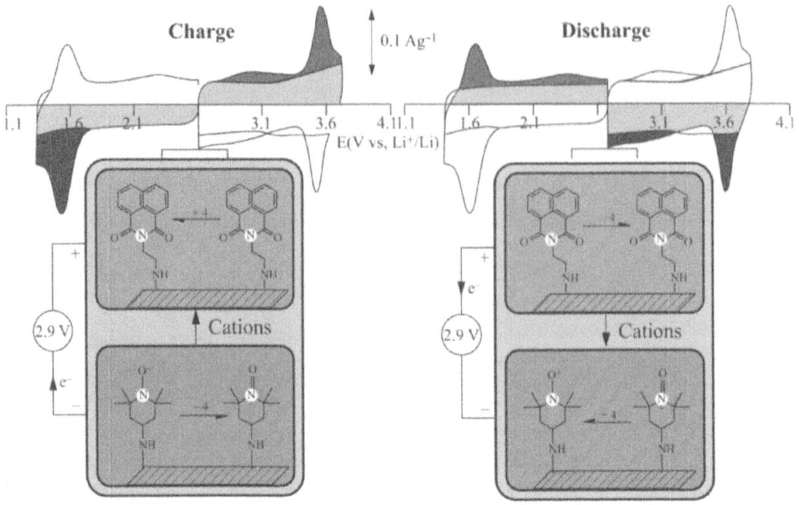

Figure 1.14. *Hybrid supercapacitors operating in an organic medium with two carbon electrodes functionalized with electroactive groups. Positive electrode: 4-amino-2,2,6,6-tetramethylpiperidinyloxy (4-amino-TEMPO); negative electrode: N-(2-aminoethyl)-1,8-naphthalimide. The cell voltage is 2.9 V and the additional capacitances associated with the presence of functional groups are represented in blue and red [LEB 14]. For a color version of the figure, see www.iste.co.uk/brousse/supercapacitors.zip*

There is, therefore, considerable room for improvement for the carbons functionalized by electroactive molecules. An important side effect is also the increase in density (in g/cm^3) of the grafted carbons that increases the volumetric energy density. The remaining challenges are improving the lifetime (cyclability) and power performances, which are the two lines of research currently being developed in laboratories.

Another approach is based on the functionalization of carbons from physico-chemical treatments leading to the formation of electroactive radicals on the surface [LIN 15]. In this case, the surface groups (quinone, carboxylic acid, etc.) are formed directly from and onto the surface carbon atoms without the addition of external molecules. Even if the method seems to be appealing due to its simple implementation, it is difficult in this case to control the formation of dedicated species over another and the control of the surface functions is much more complex to achieve than in the grafting of electroactive molecules, where the addition of electroactive molecules is controlled.

1.8. Conclusion

Porous carbon-based supercapacitors (EDLCs) are currently the most developed supercapacitors, because the electrostatic storage of the charge allows very high power (>10 kW/kg), long-term cyclability (> 1 million cycles of charge/discharge) and an operational temperature range between –40 °C and +70 °C. The main challenge EDLCs face is undoubtedly increasing their energy density (currently around 6 Wh/kg–7 Wh/L) to beyond 10 Wh/kg. From this point of view, the studies currently carried out on carbonaceous materials for use in EDLCs are taking different approaches:

– the synthesis of activated microporous carbons, with the aim of developing low-cost synthesis methods allowing a better control of the average pore size;

– the synthesis of carbons with a hierarchical porosity, to combine, for example, mesopores (electrolyte reservoir) and micropores (high capacitance). Meso-/macroporous structures may also be advantageous when using high viscosity electrolytes (ionic liquids, see section on electrolytes);

– the synthesis of porous and densified graphene, which makes it possible to achieve densities greater than 1 g/cc. Even if performance is still limited for densified graphene (cyclability in particular), recent results showed that high volumetric capacities will certainly lead to short-term applications of this type of graphene in EDLCs, besides microsupercapacitors or high-power supercapacitors;

– the synthesis of functionalized carbons, which combines carbon double-layer capacitance with surface redox capacitance. Studies carried out in aqueous electrolytes have already revealed interesting developments (e.g. using quinones); recent results showing the functionalization through the grafting of carbons with electroactive surface moieties in an organic medium is a promising step forward for grafting. However, it will still be important to preserve good cycling and power performances. Finally, Lin *et al.* [LIN 50] have recently prepared a nitrogen-doped ordered mesoporous carbon with an outstanding capacitance of 855 F/g in aqueous electrolytes that can be charged or discharged at a fast, carbon-like speed. The improvement mostly stems from pseudocapacitive redox reactions at nitrogen-associated defects that transform inert graphene-like layered carbon into an electrochemically active substance without affecting its electric conductivity:

- cross-cutting research strategy coupling in-situ experiments/modeling of transport and ion adsorption mechanisms in confined pores of porous carbons is rapidly developing, that unties renowned groups in the simulation and *in situ* characterization (electrochemistry, NMR, neutron, FT-IR, SAXS, WAXS, etc.). This research is certainly fundamental, but it also has important practical applications with regard to EDLCs or the desalination of seawater by *capacitive desalination*.

The RS2E is very active in the various research areas listed above. With laboratories such as CEMTHI (Orleans) and IS2M (Mulhouse), it holds significant skills in the field of low-cost synthesis of porous carbons, for example from biomass, with an expertise in an original technique for surface characterization of carbons – called temperature programmed desorption – coupled to a mass spectrometer. The synthesis of carbons with hierarchical porosity (micro-/mesoporous or meso-/macroporous) is explored at ICGM (Montpellier) and IS2M. Research into the functionalization of porous carbons that can increase capacitance was introduced by T. Brousse (IMN, RS2E) and D. Bélanger (UQAM) several years ago. This is an important research area at RS2E within the "capacitive storage" domain. Finally, a key line of research is the coupling of electrochemistry and modeling to understand the increase in capacitance in the nanopores and more generally the transport and the adsorption of ions in the confined pores of carbons, a field in which the RS2E is a pioneer along with the laboratories CIRIMAT (Toulouse) and PHENIX (Paris). CEMTHI, specialist in the *ex situ* RMN characterization of supercapacitor electrodes, currently develops diffusion coefficient measurements and imaging techniques that can be used for SCs.

2

Electrolytes

In this chapter, we will focus mainly on conventional electrolytes based on organic solvents and/or ionic liquids, where noticeable progress has been made over the last few years.

We will not discuss solid electrolyte or gel systems, as there has not been any significant advances in these areas in recent years, except in the case of ionogel development. The reader is invited to refer to articles in the literature for more information on these gel or solid systems [CHO 09, LEB 11a].

2.1. High potential electrolytes

With the energy density being proportional to the square of the voltage, the development of stable electrolytes at high voltages is, as in batteries, one of the most explored research topics in the field of supercapacitors.

Today, organic electrolytes mainly use are acetonitrile (AN, CH_3CN) or propylene carbonate (PC, $C_4H_6O_3$) as a solvent with fluorinated salts such as tetraethylammonium tetrafluoroborate ($(C_2H_5)_4N,BF_4$ or $TEABF_4$) (Table 2.1).

	Conductivity ($mS \cdot cm^{-1}$) RT	Viscosity ($mPa \cdot s$)
Conventional electrolytes:		
CH_3-CN (acetonitrile) + 1 M $(C_2H_5)_4N,BF_4$ ($TEABF_4$)	55	0.6
$CH_3C_2H_3O_2CO$ (PC) + 1 M $TEABF_4$	13	2.6
Another alternative organic electrolyte: $(CH_2)_4(CN)_2$ (ADN) + 0.7 M $TEABF_4$	4.3	6.6
Li ion battery electrolyte: PC:EC:DEC + 1 M $LiPF_6$ *	11.8	2.7
Ionic liquids:		
EMI,TFSI	8.4	28
PYR_{14},TFSI	2.6	62
PYR_{14},FSI,IP_{13}FSI (1:1 mass)	4.9	
Ionic liquids + solvent:		
PYR_{14},TFSI + PC (1:1 mass)	10.3	5.6
PYR_{14},TFSI + AN (1:1 mass)	57	N/A

Table 2.1. *Conductivity and viscosity of various electrolytes [BÉG 14]*

AN-based solvent electrolytes show high ionic conductivity (up to 60 $mS.cm^{-1}$ at room T), with a maximum operating voltage of 2.7 V in combination with activated carbons. PC has a similar electrochemical stability (2.7–3 V) to AN, but a conductivity about three times lower, which greatly limits the power capability of supercapacitors devices using PC-based electrolytes. The reason behind selecting PC is related to the low flash point value of AN (5°C) and the potential toxicity of the decomposition products of AN; this explains why most Japanese companies (e.g. Nipon Chemicon or Matsushita) nowadays only use PC-based electrolytes in their systems.

To increase the operating voltage without considerably compromising the power performance, several strategies have recently been developed.

The first strategy involves using ionic liquids such as salts in carbonate-based solvents. The electrolytes of Li ion batteries 1 M Li^+,PF_6^- PC:EC:DMC have been tested in supercapacitors since they offer a good compromise in terms of conductivity/stability, with maximum voltages of 3 V [LAH 11, LAH 12]. However, the presence of Li,PF_6^- anion makes them sensitive to the presence of traces of water and the formation of solid electrolyte interphase in the micropores rapidly blocks the access of ions.

In 2011, a team from the MEET Laboratory in Munster proposed a $PYR_{14}TFSI$ 2 M electrolyte in PC only [KRA 11a].

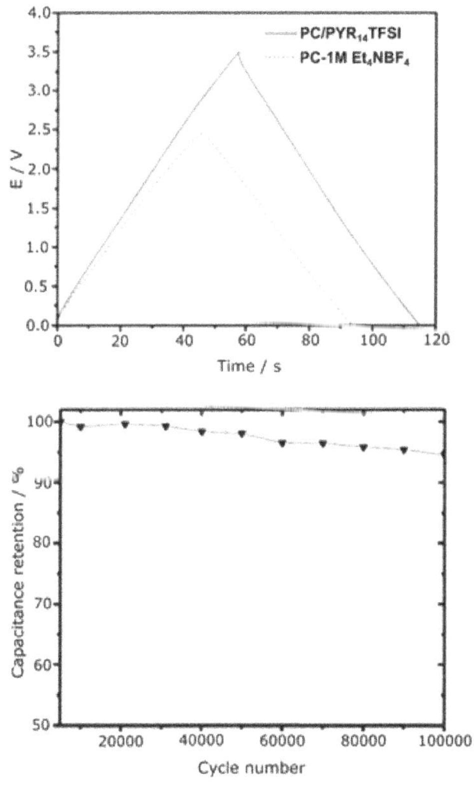

Figure 2.1. *Galvanostatic cycling curves of an activated carbon based cell (Norit carbon) in a $PYR_{14}TFSI$ 2 M electrolyte in the PC [KRA 11a]*

The maximum operating voltage is 3.5 V, with remarkable cycle stability (–5% over 100,000 cycles). These results are quite surprising considering that electrolytes based on PC and conventional fluorinated salts such as $TEABF_4$ decompose into propylene, CO, H_2 and CO_2 above 3 V [SHU 09].

Koetz *et al.* [WEI 13] proposed an explanation by comparing the degradation of cycling performance of supercapacitors using two different tests: galvanostatic cycling and floating. They have found that galvanostatic cycling does not lead to reliable information regarding the electrochemical stabilities of the electrolytes since the cumulative holding time at high voltages during this type of cycling is short compared to the total cycling time. This could explain the results obtained by Krause [KRA 11a]. However, another explanation of this improvement could be a better electrochemical stability of the reduced pyrrolidinium cation compared to conventional TEA^+ salts.

Another approach is the use of new solvents. This approach is potentially the most interesting, but also the most complicated because of the multiple criteria that candidates must meet (conductivity, electrochemical stability, viscosity, etc.). However, there are some interesting studies including those developed by the MEET Laboratory (Germany) with adiponitrile (ADP) [BRA 12a, BRA 12b].

The limitations of using ADP as a solvent are i) its moderate conductivity (Table 2.1) and ii) high viscosity, which limit the capacitance of porous carbon at around 40 F/g. As its conductivity greatly decreases as temperature decreases, ADP cannot be used below 0 °C. Because of these limitations, ADP may only be an attractive solvent for some specific applications.

Another strategy is to use pure ionic liquids to take advantage of their much greater stability window compared to other electrolytes. Another approach has been to add a pseudocapacitive contribution to the carbon double-layer capacity, as observed with acid electrolytes. For this purpose, protic ionic liquids (PILs) have been studied (Figure 2.2).

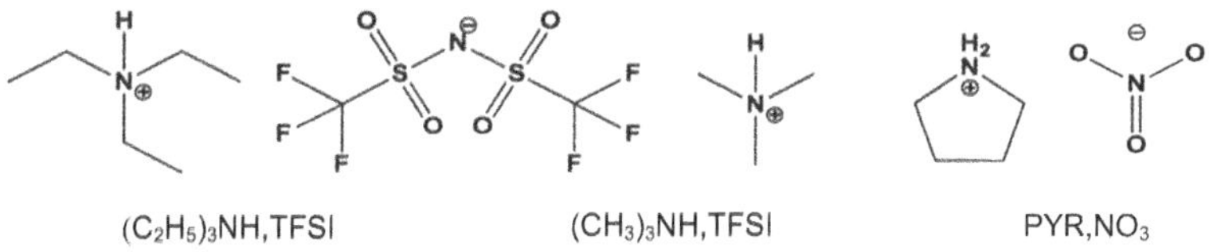

Figure 2.2. *Protic ionic liquids [BRA 13, BOI 13]*

PILs have the advantage of being less expensive and less viscous than aprotic ILs, but the presence of protons limits the potential window due to the possible reduction of protons, resulting in a maximum voltage of 2 V [BOI 13, MEN 13]. Moreover, the capacitance of porous carbons in PILs remains limited due to the viscosity which, although less than that of the aprotic IL, hinders access to the all the pores of the activated carbons. Another disadvantage is the decrease in conductivity and hence a decrease in performance with decreasing temperature, which again restricts the use of PILs at positive temperatures. Research is oriented toward the design of stable high-voltage PILs, but even in this case, controlling the water content is also a major challenge to overcome before using these systems.

An alternative approach to tackling these issues is to use pure aprotic ionic liquids to essentially exploit the wide electrochemical stability window. However, there are two major problems: the high viscosity of the ionic liquids (Table 2.1), which limits the accessibility to the carbon pores (limited capacitance), and the limited conductivity below room temperature.

An original approach considered within the RS2E involves developing specific carbon/electrolyte interfaces to maintain maximum pore accessibility and allow operation at low temperatures by maintaining a high potential window (>3 V).

To keep the surface of the carbon highly accessible to the electrolyte, the idea is to use carbons with a high surface area, but are fully accessible such as carbon onions (OLCs) or activated graphene [LIN 11, TSA 13b].

The stability of conventional electrolyte at high potential (>3 V) is an issue, as well as its operation ionic liquids at temperatures <0° C; these can be overcome by using ionic liquid mixtures [LIN 11, TSA 13b]. Figure 2.3 shows the change in conductivity versus the reciprocal of the temperature for two ionic liquids (PYR_{14},FSI and PIP_{13},FSI) and a 50% mixture by mass of the two ionic liquids. The melting points are +6 and –18 °C for PIP_{13},FSI and PYR_{14},FSI, respectively. The change in conductivity of the mixture (Figure 2.3(a)) shows, on the one hand, the absence of any phase transition (melting or solidification) down to –80 °C and, on the other hand, a decent conductivity of the mixture of 4 mS/cm at room temperature. This result can be explained by fewer interactions between the ions.

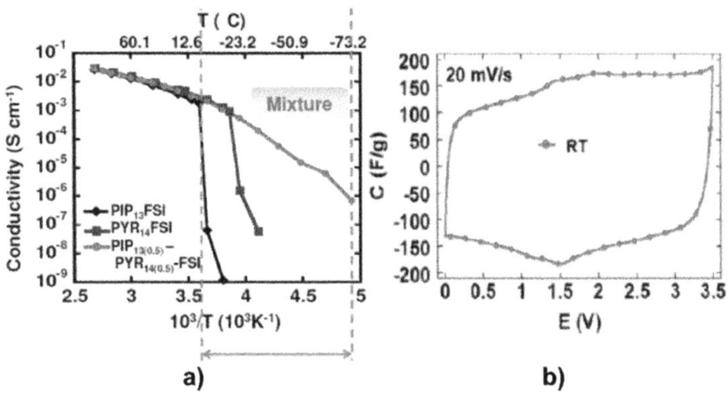

Figure 2.3. *a) Conductivity of ionic liquids PYR_{14},FSI and PIP_{13},FSI, as well as that of a 50 wt% PYR_{14},FSI:PIP_{13},FSI mixture; b) cyclic voltammetry of a cell assembled with 2 mg/cm² of activated graphene in the mixture of ILs at RT [LIN 11, TSA 13b]. For a color version of the figure, see www.iste.co.uk/brousse/supercapacitors. zip*

The mixture contains the same anions (FSI^-) and dissymmetric cations with the same molar weight, which decreases the interactions between ions. As a result, the mixture conductivity plot highlights the absence of solidification at low temperatures: the viscosity is still high, but the conductivity decreases continuously, confirming the absence of a state change.

This electrolyte has been tested in supercapacitors not using a conventional porous carbon, but a porous carbon with a fully accessible surface: activated graphene. The operating voltage is 3.7 V at 25 °C, with capacitance greater than 170 F/g. It doubles the energy density compared to an AN-$TEABF_4$ electrolyte operating at 2.7 V; the capacitance is greater than that of standard microporous activated carbon (Kuraray YP50 or YP17). Figure 2.4 shows CVs at –40 and –50 °C, i.e. at a temperature at which the AN-$TEABF_4$ electrolyte is solid and cannot function. Even if the potential scan rate remains modest (1 mV/s), the systems can operate at these extreme temperatures, which is a record for an electrolyte based on pure ionic liquids. Experimental characterizations have shown that the operation temperature can be pushed up to 80 °C and even 100 °C [LIN 11], which is the widest operating temperature range ever reported for an electrolyte (Figure 2.4).

These results validate the approach that involves developing the carbon structure (porous structure) in conjunction with the electrolyte properties (viscosity in particular). Other combinations of ionic liquids are also possible based on the same principle (dissymmetric cations and same anions, such as with PYR_{23},FSI:PYR_{14},FSI). The wide variety of available porous carbon structures, resulting from the new syntheses presented in Chapter 1 and in particular hierarchically porous carbons (section I.4), offers multiple carbon/electrolyte combinations to develop systems meeting the needs in terms of power, energy and temperature requirements.

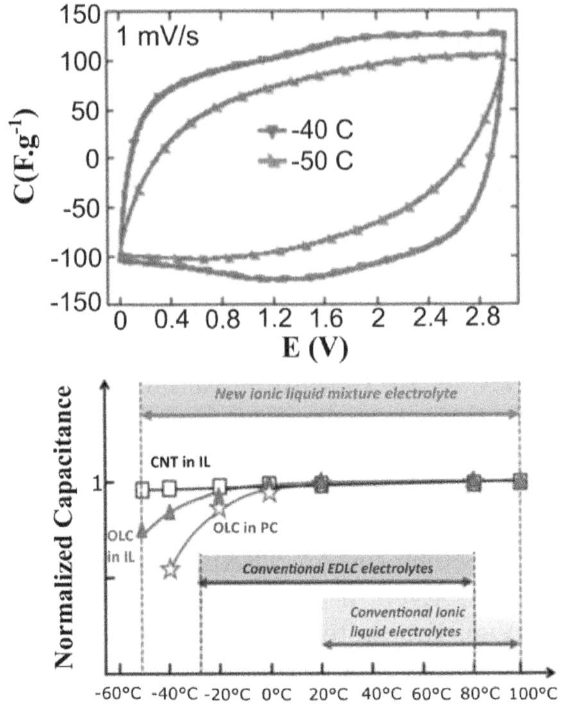

Figure 2.4. *a) Cyclic voltammetry of supercapacitors assembled with activated graphene in a mixture of ionic liquids PYR_{14},FSI and PIP_{13},FSI at 40 and –50 C; b) operation temperature range of the electrolyte based on a mixture of ILs in comparison with standard electrolytes (PC- or AN-$TEABF_4$) and pure ionic liquids [LIN 11, TSA 13b]. For a color version of the figure, see www.iste.co.uk/ brousse/supercapacitors.zip*

2.2. What about AN?

AN is the electrolyte solvent currently leading the way to the best electrochemical performance in terms of power densities (the purpose of SCs) and offers excellent energy densities (about 6 Wh/kg). Today, the alternative to AN in commercial products is PC (or carbonate mixtures), which offers similar cell voltages but power densities three times lower. AN[1] (CAS number: 75-05-8, molar weight: 41.05 g·mol^{-1}) is harmful if it is inhaled, in contact with skin, or if swallowed. It is also an eye irritant. Furthermore, it has two main disadvantages regarding applications in electrochemical storage:

– first of all, it is a volatile solvent with a boiling point at 81 °C and a flash point at 5 °C, with a risk of flammability;

– also, it contains a –CN functional group, causing fear of HCN formation during its eventual decomposition (under an uncontrolled voltage or during cell combustion).

What about its current use in supercapacitors?

One explanation is found in the analysis of commercial cells. Obviously, no Japanese manufacturer uses AN in supercapacitors marketed in Japan; these manufacturers, such as Nipon Chemicon, prefer PC, despite the limitations induced in the power performance of the devices. However, the company NessCap (Korean) offers two product ranges: those based on PC for cells to be commercialized in Japan and those based on AN for the rest of the world. Blue Solution (Bolloré) only sells SCs based on AN for large-format cells (>2,000 F) in electric traction applications (streetcar, boat, bus, etc.), which are currently still only used in Europe. The Maxwell company, another pioneer company that commercializes products based on AN, announced the opening of a SC manufacturing plant in Japan during the AABC conference (June 2013), dedicated to the Japanese market. The solution considered to continue to use AN involves confining the electrolyte to avoid any leakage, which is manageable (as in batteries, for example). Nonetheless, replacing AN is currently an important research topic.

1 See: www.chups.jussieu.fr/recherche/fichesdesecuriteproduits/acetonitrile trespur.pdf.

2.3. Conclusion

The electrolytes used in commercially available SCs currently contain AN or polypropylene carbonate (PC) as the solvent. The low conductivity of PC and its viscosity limit the power of systems, yet AN still raises some questions about its use from a safety point of view (release of potentially toxic gases). As neither of these two solvents are without their flaws, AN stands out today because of its performance and the absence of problems to date when used in many applications.

Despite research on electrolytes being oriented in different directions, all seek the same objective: to increase the electrochemical stability window in order to increase the operating voltage and therefore the energy.

Some strategies include:

– the search for alternative organic solvents. Even if "new" solvents have been proposed (ADP, DMSO, polysulfones, etc.), it is difficult to develop an organic compound matching with all the criteria of the ideal electrolyte (low viscosity, high conductivity, electrochemical stability, etc.). The improvements are only incremental, but this field of research remains active;

– the use of ionic liquids as salts in conventional solvents such as PC. If the potential window is increased due to the greater electrochemical stability of the pyrrolidinium cations, this does not solve the drawback of cost or limited power;

– the use of aprotic ionic liquids to combine double-layer capacitance with pseudocapacity has not yet generated spectacular results. The limitations to overcome are: the viscosity, which causes a significant reduction in the capacitance of the carbons; the conductivity below ambient temperature; the control of the water content (window of electrochemical stability <2 V). This research is developed within the RS2E;

– the control of the carbon structure according to the nature of the electrolyte came out with cells operating in a T range of 140 °C, at 3.7 V at room temperature, with capacitances of 170 F/g for carbon (activated graphene) in mixtures of ionic liquids. This research direction seems promising and the possibility of controlling pore size distributions with template or hierarchical porosity carbons offers a large number of possible combinations. This work is being developed in the RS2E.

For the latter two approaches, a challenge will be to make them compatible with the cost criteria.

Gel electrolytes based on ionic liquids are currently flourishing due to their applicability to microsupercapacitors. They are also a promising way of increasing the safety of massive devices, in particular with regard to their use at high temperatures [BRA 14, BRA 16].

Finally, it is worth to mention here that another approach has very recently been published by a group at the University of Montpellier (O. Fontaine and F. Favier) within the RS2E. It consists of grafting an active redox group onto ionic liquids salts (called biredox ionic liquids) to combine both a capacitive storage by ionic liquid adsorption into carbon pores and a Faradic storage by redox reactions of the functionalized ions [MOU 16]. The use in supercapacitors represents a first demonstration of the unique possibilities of biredox ionic liquids, which can more generally enable high-capacity/high-rate charge storage.

3

Pseudocapacitive Materials

The prefix *pseudo* comes from the ancient Greek *pseudês*, which means "false" or "erroneous". This prefix can give two meanings to the word it precedes[1]: it indicates that the quality they express is false, or that it does not suit the thing or the person (e.g. a pseudoscientist, who is a person wrongly mistaken for a scientist); it can also mean "that looks like" but is of a different nature (e.g. a pseudoword that is a string of characters that resembles a real word but has no meaning. Pseudowords respect the orthographic conventions of a language and are therefore pronounceable, like words).

With regard to the term "pseudocapacitive", the second definition applies [BRO 15]. It should therefore be understood that this term can only refer to electrode materials that have most of the characteristics of conventional capacitive electrodes (capacitance originating from the electrochemical double layer), the main one being a linear change in the stored charge Q (in C) with the potential ΔE (in V). This linearity is expressed by means of the capacitance C, expressed in F in the following relation:

$$Q = C \times \Delta E \qquad [3.1]$$

1 *Académie française Dictionnary*, 1932–1935, 8th edition.

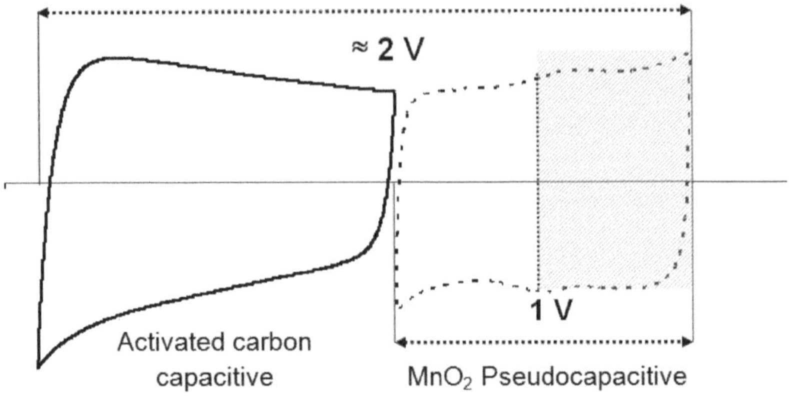

Figure 3.1. *Cyclic voltammogram (CV) of an asymmetric carbon/MnO_2 device in an aqueous electrolyte containing 0.5 M K_2SO_4. Note that the shape of the two CVs are quite similar, despite the activated carbon being a capacitive material and manganese dioxide being a pseudocapacitive one [BÉG 13]*

This main feature of pseudocapacitive electrodes has unfortunately been forgotten over the past decade for various reasons, the main one being a misunderstanding of the composition of hybrid devices in which the battery electrode is confusingly likened to a pseudocapacitive electrode. This caused the term pseudocapacitive to be assigned to a large number of materials, which are purely Faradaic electrode materials such as those used in batteries, the most striking example being nickel hydroxide $Ni(OH)_2$. We will return to this concept in the following sections. It is therefore useful to understand the choice of materials presented below, which correspond to the definition given above.

Pseudocapacitive materials involve rapid and reversible redox reactions responsible for charge storage. Compared to ionic adsorption in EDLCs, these Faradaic contributions substantially increase the resulting capacitance and thus the energy of the device. However, the gain in energy is achieved detrimentally to the power performance. Indeed, the redox reactions involved, although rapid, have slower kinetics than that of ionic adsorption phenomena as the latter do not involve any charge transfer or ionic diffusion within a solid, or modification of the crystallographic structure of the electrode material. In pseudocapacitive materials, the electrostatic contribution of the double layer to the total capacitance is generally low due to limited electronic conductivity and/or a moderate surface area of the materials.

The most studied pseudocapacitive materials to date are electronic conductive polymers and metal oxides. The performance of transition metal nitrides has also been explored. Depending on the redox couples involved, these materials are either considered as positive or negative electrodes. There are many possible combinations of these various electrode materials for the design and fabrication of devices with specific performances. The large number of formulations of organic or (more generally) aqueous electrolytes adds to the array of technological possibilities.

The reference compound that allowed Conway to demonstrate pseudocapacitive storage is hydrated ruthenium oxide [CON 91]. RuO_2 is a non-stoichiometric oxide with Ru showing various valencies (from +II to +IV). It shows metal-like electrical conductivity and excellent ionic conductivity based on solid state proton mobility. The cyclic voltammetry in an acidic (Figure 3.2) or alkaline electrolyte leads to a capacitive signature similar to that obtained with EDLCs: anodic and cathodic scans are almost overlapped (*mirror image* such as proposed by Conway) with an almost constant charge (and discharge) current over a window of 1.4 V. This CV can actually be described as successive perfectly reversible redox reactions over 1.4 V [CON 13]. The charge storage mechanism is essentially Faradaic and involves an equilibrium in which the charge balance is achieved by diffusing protons in the hydrated oxide matrix coupled with electron exchange:

$$RuO_x(OH)_y + \delta e^- + \delta H^+ \leftrightarrow RuO_{x-\delta}(OH)_{y+\delta} \quad [3.2]$$

Most of the Faradaic reactions here take place at the surface – or in the immediate vicinity of the surface – and are therefore not limited by diffusion [ARD 90].

Despite specific capacities of 450–700 F/g over several tens of thousands of charge–discharge cycles, the toxicity and the cost of RuO_2 limit its use to niche industrial applications (e.g. microdevices) but remains, from a more academic point of view, as the pseudocapacitive model material.

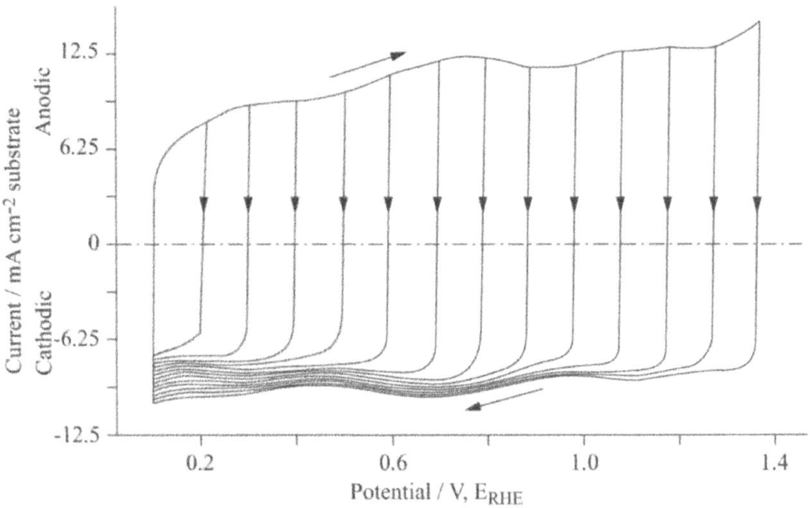

Figure 3.2. *Cyclic voltammogram of a RuO_2 electrode in H_2SO_4 (1 M) [CON 13]*

30 years after Conway defined the concept of pseudocapacitance, and with the first commercial pseudocapacitors soon to arrive on the market, the initial specifications and definitions for pseudocapacitance remain valid: multiple oxidation states within a large potential window, good ion exchange properties, high electronic conductivity and so on. Like Faradaic materials used in batteries, but unlike EDLCs, the electronic and crystalline structures as well as the microstructures of the pseudocapacitive materials can be quite different from one material to another. They are generally strongly correlated with the charge/discharge mechanism and associated performances. As a matter of fact, current research in the field is mainly based on the triptych: "formulation/(nano) structuration, storage mechanism, performance".

Many materials, usually metal oxides, are still too often presented in the literature as pseudocapacitive materials [BRO 15] on the basis of rather extravagant specific capacitances, originating from their sampling in the form of thin films (a few nanometers for a few nanograms). These materials are not relevant since (1) they are not really pseudocapacitive and (2) the areal capacitance (few tens of µF per cm^2 of electrode) is too limited to be used in real-life large-size devices. Given the importance of supercapacitors in recent years, it is mandatory to establish a clear classification to avoid overselling or hype, or biased announcements such as recently seen with the Intel Award Prix Intel[2] (a supercapacitor cannot replace a battery to power a smartphone, unless one wants to use it for no more than 30 s). In this context, the RS2E coauthored a recent article in the *Science* journal, which defines the notion of intrinsic capacitive behavior: the capacitive signature does

2 See: http://phys.org/news/2013-05-high-school-student-supercapacitor-young.html.

not depend on the implementation (thin layer) or the composition (doping) [SIM 14].

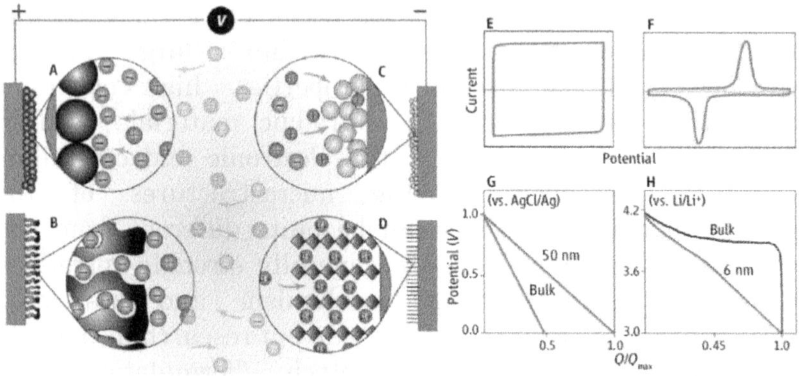

Figure 3.3. *Comparison of batteries and supercapacitors [SIM 14]. For a color version of the figure, see www.iste.co.uk/brousse/supercapacitors.zip*

NOTE.– Figure 3.3 shows the different capacitive storage mechanisms, with the charging of the electrochemical double layer on solid (a) or porous (b) carbon electrodes. The pseudocapacitive mechanisms include the mechanism of redox pseudocapacitance (c) as with MnO_2 or RuO_2 and the peudocapacitive intercalation reaction observed during the intercalation of Li+ ions in Nb_2O_5 (d), (e) and (h). Various electrochemical signatures which makes it possible to distinguish between supercapacitors and batteries are shown. The capacitive signature in cyclic voltammetry test results in a constant current during constant potential scan rate (e) while that of a battery shows redox peaks (f). The galvanostatic discharge of a pseudocapacitive electrode of MnO_2 is linear (g), whatever the particle size (massive or nanometric), whereas the signature of a $LiCoO_2$ battery electrode depends on its morphology (h): linear (*pseudocapacitive-like*) for nanometric particles and battery-like for bulk (microsized) particles.

3.1. Conductive polymers

The charge storage mechanism in electrically conducting polymer (ECP) materials is based on the creation of structural defects (polarons and bipolarons) resulting from the oxidation (or reduction) of the polymer chain [SCR 95]. The term p or n doping refers to the oxidation or reduction of a neutral polymer, respectively. Several configurations are possible when designing devices:

– two electrodes using the same p-doped polymer (type I);

– a different p-doped polymer at each electrode (type II);

– an n-doped polymer at the negative electrode and a p-doped polymer at the positive electrode (type III).

System (type)	Electrolyte	Deposit	Voltage (V)	Energy (Wh/kg)	Power (W/kg)
PPy/PPy (I)	Organic	Chemical	1.2	~4	~50
PEDOT/PEDOT (I)	Organic	Electrochem.	0.8	1–4	35–1,400
PANI/PANI (I)	Organic	Electrochem.	1	3–4	250–1, 400
PANI/PANI (I)	Aqueous	Electrochem.	0.75	2–3	~1,000
PPy/PANI (II)	Organic	Electrochem.	1	4–5	250–1,200
PFPT/PFPT (III)	Organic	Electrochem.	3	25–40	2,000–10,000

Table 3.1. *Supercapacitors based on conductive polymers and their main characteristics [LAF 01b]*

Apart from Type III systems, performance is quite limited in terms of voltage, energy density and power. Along with these disadvantages, the synthesis costs are still too high and the number of charge–discharge cycles is limited (<10,000) due to the mechanical degradation of the electrode by loss of material and trapping of the polarons. Therefore, industrial perspectives are limited with this type of device [SNO 11].

Improvement strategies have been explored. Several groups have thus used p-doped polythiophenes positive electrodes associated with activated carbon-based negative electrodes in hybrid devices with fewer cycling losses [VIL 03, LAF 01a]. Various nanocomposites have also been proposed, including pseudocapacitive oxides [LU 11], carbon nanofibers, nanotubes [WAN 11] or graphene [WU 10]. Although there has been a marked increase in the number of research studies in this field over the last 3 years, the improvements observed are still insufficient and the commercialization of a device based on ECPs is still uncertain.

3.2. Metal oxides

Metal oxides are still the most studied pseudocapacitive materials [BRO 15] and probably the materials with the greatest potential for industrial development. For this reason, they must meet certain specifications: a simple synthesis process from low-cost materials, good power and energy densities (because of participating Faradaic reactions).

Manganese dioxide, MnO_2, is, in this respect, the pseudocapacitive material of choice [BRO 15]. It is naturally abundant, inexpensive and not too toxic. Its charge storage mechanism has been described as involving very fast Mn^{3+}/Mn^{4+} redox reactions with cation exchanges, protons and cations of the electrolyte, to balance the change in the oxidation state of manganese [TOU 04]:

$$MnO_2 + (x + y)e^- + xH^+ + yC^+ \Rightarrow MnOOH_xC_y \qquad [3.3]$$

where C^+ is the cation from the electrolyte.

The literature reports capacitances of approximately 120–250 F/g and cycling has been demonstrated over several hundred thousand charge/discharge cycles, usually lasting several dozen seconds [BRO 07]. The energy density is greater than 5 Wh/kg. It operates in aqueous electrolytes (up to temperatures of –20 °C) and is therefore safer and green.

In situ X-ray diffraction measurements at various charge/discharge states have shown that the charge storage mechanism, which depends on the variety of MnO_2 considered, also involves the bulk of the electrode material [GHO 09]. The evidence of the pseudocapacitive bulk mechanism and the associated electrochemical performance lies in the remarkable cation exchange capacity of the manganese dioxides and are generally insensitive to an increase in the active surface area developed by the material. This is called "intrinsic pseudocapacitance" [AUG 14].

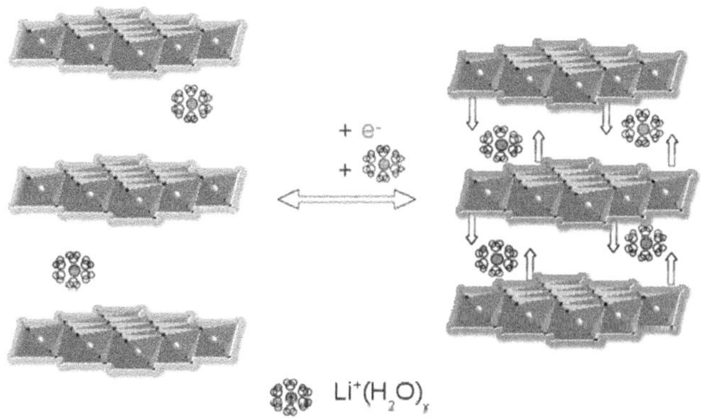

Figure 3.4. *Electrostatic breathing of the MnO_2 birnessite layered structure under the reversible intercalation of hydrated cations of the electrolyte [GHO 12]. For a color version of the figure, see www.iste.co.uk/brousse/supercapacitors.zip*

The power performances of pseudocapacitive oxides, in particular manganese dioxides, are also limited by their low electronic conductivity. This is not the case for RuO_2 and the performances of other oxides, particularly in the spinel form, have also been explored: Mn_3O_4, Fe_3O_4, Co_3O_4, $MnFe_2O_4$, $NiCo_2O_4$, etc. For the last three, their storage mechanisms are Faradaic and not pseudocapacitive, and are related to their reactivity toward the electrolyte, especially when strongly alkaline [AUG 14]. $Ni(OH)_2$ [XIN 11] and $Co(OH)_2$ [ZHO 09b] lamellar hydroxides are sometimes described as pseudocapacitive, but they are, in fact, clearly battery-type electrode materials [BRO 15].

For the oxides presented above, electrolytes are aqueous, alkaline, acidic or neutral, depending on the stability of the materials considered and the charge storage mechanisms involved. The energy of the system is then also limited by the stability of the electrolyte, usually over a potential window of 1.5–2 V maximum. Some pseudocapacitive oxides can also be used in an organic electrolyte, with a composition similar to that used with Li ion batteries. Its advantage, due to a much wider voltage window, is obvious from an energy point of view. This is the case for V_2O_5, whose layered structure gives rise to a charge/discharge mechanism, which can be controlled by both intercalation and pseudocapacitive (extrinsic) phenomena when the material is in the form of a nanocomposite associated with a nanostructured carbon (e.g. carbon nanotubes) [SAT 11]. Pseudocapacitive behavior by intercalation (of Li^+) in the layered structure of TiO_2(B) [ZUK 05] and the orthorhombic form of T-Nb_2O_5 [AUG 13] have also been assessed.

Current research on pseudocapacitive materials begins with the elucidation of the storage mechanisms, specific to each material. With performance goals and therefore more practical and technological goals, it also aims to increase (i) the power density of pseudocapacitive materials through the association with conductive compounds with large surface areas and (ii) the volumetric energy densities by choosing electroactive materials with greater mass density.

Decorating large specific surface area carbons with pseudocapacitive nanooxides is a strategy that leads to an increase in the power density. The aim of this approach is to maximize the electronic percolation and the developed active surface area. Various nano- or mesostructured carbons can be used. The precipitation or the electrodeposition of oxides in porous structures is preferentially achieved with carbons showing an interconnected network of meso- or macropores, which are large enough to allow ion accessibility and easy transport of electrolyte ions throughout the whole volume of the electrode. However, with these carbons (non-microporous), the gain in surface area and therefore in capacity are unfortunately limited. Carbons with open structures, such as graphene (two-dimensional) or nanotubes (1D) and nanofibers (1D), are thus preferred. Carbonaceous 2D and 1D materials offer many opportunities for designing/shaping electrodes, and by extension, flexible devices. Numerous publications report "remarkable" performances, especially in terms of power, with metal oxides and hydroxides on graphene such as, for example, with MnO_2 [CHE 10], Co_3O_4 [YAN 10] or $Ni(OH)_2$ [WAN 10]. However, for the latter two, their electrochemical signature is difficult to compare to that of a pseudocapacitive material.

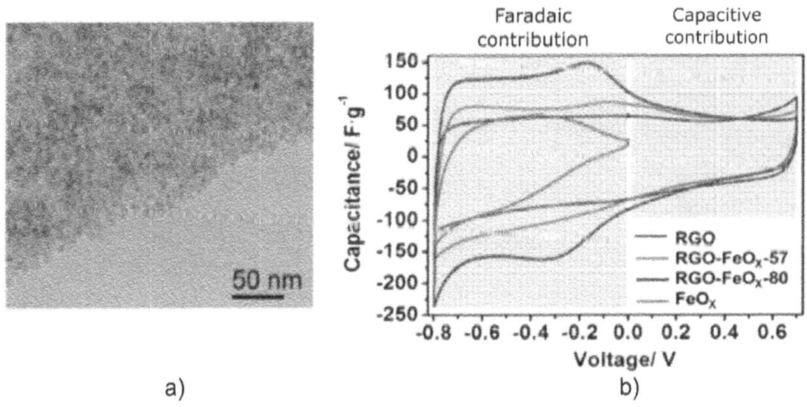

Figure 3.5. *a) Scanning electron microscopy (SEM) image of FeOx nanoparticles on RGO; b) voltammograms of electrodes based on FeOx, RGO and composites (Na2SO4, 1 M and voltage relative to Ag/AgCl) [GAO 14]. For a color version of the figure, see www.iste.co.uk/brousse/super capacitors.zip*

The microwave reduction in benzyl alcohol of graphitic oxide (GO) exfoliated in the presence of iron (III) acetylacetonate leads to the deposition of a dense layer of nanoparticles of FeO_x iron oxide (mixture of magnetite and maghemite) on reduced GO sheets (RGO). As shown in Figure 3.5, the materials obtained have a combined behavior depending on the potential window considered, which is clearly pseudocapacitive in the negative part (–0.8 to 0 V versus Ag/AgCl), where the iron oxide is active, and rather capacitive in the positive window (0–0.7 V versus Ag/AgCl). In the extended potential window of 1.5 V, the measured capacitances for the nanocomposites reach 100 F/g with a power behavior similar to that of pure RGO [GAO 14].

Decorating carbon nanotubes or nanofibers with MnO_2 is a well-known approach of RS2E partners [KHO 05, GHO 13]. In addition to the attractive possibilities of manufacturing flexible electrodes, nanotubes or nanofibers provide electronic continuity ensuring capacitive performances of approximately 140 F/g over an extended potential window of 2 V in a neutral aqueous electrolyte and sustained charge/discharge regimes.

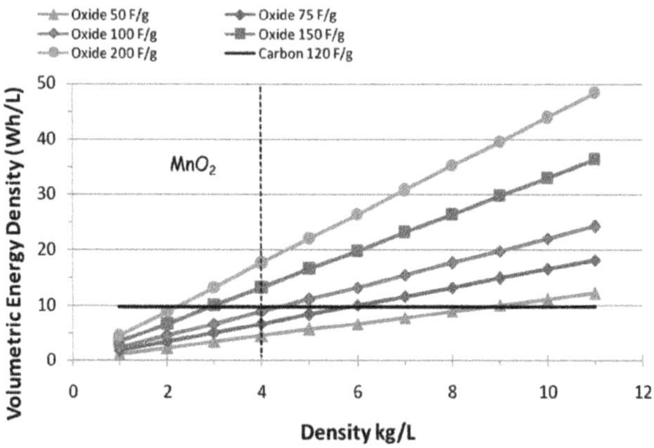

Figure 3.6. *Volumetric energy density of various oxides according to their mass density [GOU 16]. For a color version of the figure, see www.iste.co.uk/brousse/supercapacitors.zip*

For many stationary or automotive applications, the size, and therefore volume of devices, takes precedence over mass consideration. Figure 3.6 shows the change in the energy density of oxides with different specific capacitances (in F/g) according to their density [GOU 16]. The energy density at 10 Wh/L of a state-of-the-art activated carbon at 120 F/g is given for comparison. With a density of 4 kg/L, a manganese dioxide at 150 F/g (which is a standard value for this material) already exceeds this threshold and obviously oxides with greater densities, is even more advantageous, since a capacitance of 50 F/g for an oxide of 10 kg/L is sufficient to exceed the performance of carbon.

3.3. Transition metal nitrides

Recently, metal nitrides such as Mo_xN [CON 98, GAO 13a], TiN [CHO 05], VN [CHO 06], WN [KAR 12] and RuN [BOU 13] have been studied. Impressive specific capacitance values up to 1,340 Fg^{-1} have been reported for VN [CHO 06] thin layers.

The CV of a thin film of Mo_xN is shown in Figure 3.7(a). The pseudocapacitive nature of this material is confirmed in an acidic medium since a linear variation in the charge is observed within the explored potential range. For such an electrode, it is easy to determine a capacitance (in F) in the same way as with a carbon-based electrode, even if the phenomena governing the charge storage are different.

On the other hand, a pseudocapacitive character is more difficult to confirm on electrodes based on vanadium nitride (Figure 3.7(b)) where, apart from the pseudocapacitive envelope, redox peaks are also observed.

According to Choi *et al.*, the impressive specific capacity of VN is due to its high electronic conductivity (σ_{powder} = 8.82 × 103 Ω^{-1} m^{-1}) associated with fast and reversible surface redox reactions [CHO 06]. The high capacitances result from the successive redox reactions that occur due to the partially oxidized surface of VN (VN_xO_y) in the presence of electrolytic OH^- ions, as well as the capacitance of the double layer, to a lesser extent.

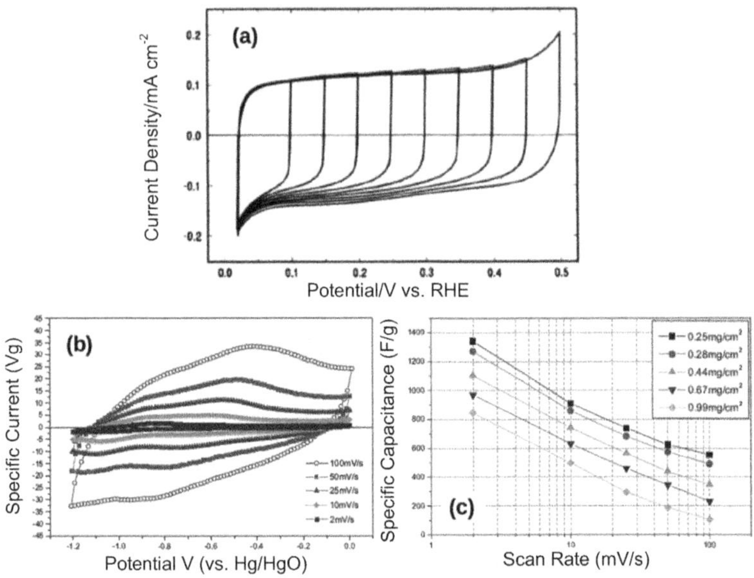

Figure 3.7. *Cyclic voltammetry for: a) thin films of Mo_xN in H_2SO_4 0.5 M [CON 98]; b) VN nanocristals synthetized at 400 °C [CHO 06] at various scan rates (2–100 mVs^{-1}) in KOH 1 M; c) specific capacity for various weights of VN according to the scan rate [CHO 06]*

The bulk of the VN film electrode allows the rapid transport of electrons and consequently a very rapid charge percolation. The following reaction has been proposed to illustrate the whole process [CHO 06, PAN 12]:

$$VN_xO_y + OH^- \rightarrow zVN_xO_y//OH^- + (1-z)VN_xO_y\text{-}OH + (1-z)e^- \quad [3.4]$$

$VN_xO_y//OH^-$ represents the charge of the double layer capacitance and VN_xO_y-OH is associated with the reduction of the vanadium cations initially present in the compound.

However, significantly lower specific capacitance values for VN (161 Fg^{-1}) have been found in different reports [ZHO 09a, GLU 10, LUC 12]. Such a discrepancy between these values and those reported by Choi [CHO 06] may be partly due to the difference in the oxygen content in the pristine material, and/or to the difference in the average oxidation state of the vanadium cation at the material surface [CHO 06, ZHO 09a, GLU 10, LUC 12]. These preliminary reports suggest that the specific capacitance of transition metal nitrides is controlled by the surface composition. Thus, appropriate synthesis methods are needed to control the surface composition.

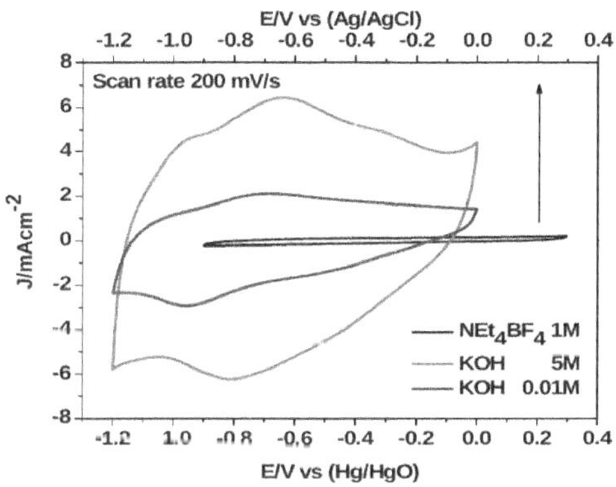

Figure 3.8. *Cyclic voltammograms of a thin VN film with a thickness of 480 nm (scan rate = 200 mV/s) in different concentrations of KOH over 100 cycles. The voltammogram of the same electrode in 1 M NEt_4BF_4 in acetonitrile is given for comparison [ACH 14]. For a color version of the figure, see www.iste.co.uk/brousse/supercapacitors.zip*

Vanadium nitride has been prepared by various routes such as nitridation of oxides, ammonolysis of metals, from various nitrogen sources such as cyanamides, urea or melamine [GLU 10, LUC 12, BUH 07, GIO 09, TES 03, SHU 13, GAO 12b, CHE 11]. It has also been obtained by the metathesis reaction in the solid state using metal oxides as precursors and metal azides as nitrogen source. Such processes have also been used to prepare carbon nanotubes coated with VN [XIA 13, ZHA 11, GHI 11]. Other approaches use TiN nanotubes as a substrate for the preparation of TiN/VN core–shell composites [ZHO 11, DON 11]. For all these approaches, high temperatures and long-duration synthesis are needed [SON 07, HEC 98].

To overcome these limitations, physical methods have also been preferred, such as thin film deposition of VN by pulsed laser and reactive magnetron sputtering. The deposition of thin films of VN allows to control (and limit) the oxygen content, but also avoid the need for binders and conductive additives such as in composite electrodes, which are detrimental when studying the intrinsic properties of the material. Moreover, due to their high electronic conductivity, the VN thin film electrodes can be used both as the current collector and as an electrochemically active material (working electrode) [PIE 14, ACH 14, LU 13, EUS 13, GAO 13c].

From an economic point of view, transition metal nitrides are now commonly used in tools such as drills, for example, and produced at very low costs. The main challenge lies in moving from thin films with attractive performances to bulk materials in the powder form, which can be used directly in electrodes of sizeable devices, in terms of energy density.

3.4. Conclusion

Due to the Faradaic reactions involved in the charge storage, the energy densities of pseudocapacitive materials are greater than those of EDLCs. *A contrario*, the corresponding power densities are lower because of the substantially slower kinetics. If this research aims to elucidate the storage mechanisms, it also involves the research of new materials, their (nano)structuration, their formulation as electrodes and their integration into devices:

– the synthesis of composites combining a pseudocapacitive oxide with a meso- or nanostructured carbon to increase the electronic conductivity and power;

– the synthesis of high-density materials to increase the energy density and polycationic oxides such as $FeWO_4$ [GOU 15];

– the synthesis of metallic nitrides with attractive performances, in particular in the form of thin films.

The cyclability of the pseudocapacitive materials/ electrodes is also a key point that must be validated for the development of devices incorporating such electrodes.

4

Hybrid and/or Asymmetric Systems

Carbon-based EDLCs operated in non-aqueous electrolytes show higher energy and power densities due to the larger potential stability window (2.5–2.8 V) of organic electrolytes than that of aqueous ones (<1.5 V). The lower capacitance of carbon electrodes in organic electrolytes (vs. aqueous) is balanced by the increase in potential and subsequently in cell voltage, since the maximum energy E_{max} increases with the square of the maximum operating voltage (U_{max}). Therefore, considering that the capacitance of an activated carbon electrode in an organic electrolyte is half of that in an aqueous one ($C_{aq} = 2\ C_{org}$), and that the maximum cell voltage in an organic electrolyte is twice that in an aqueous electrolyte ($U_{org} = 2.7V \approx 2\ U_{aq}$), the relationship between the energy E_{org} in organic electrolyte and E_{aq} in aqueous electrolyte is:

$$E_{org} = \frac{1}{2}\ C_{org}\ U_{org}^2 = \frac{1}{2}\ (\frac{1}{2}\ C_{aq})\ (2\ U_{aq})^2 = 2\ E_{aq} \qquad [4.1]$$

Experimental determinations of energy confirm the better performance of EDLC systems in organic electrolytes compared to aqueous electrolytes (6 Wh/kg in organic compared to 2 Wh/kg in an aqueous electrolyte) [SIM 08b]. This has been proven practically and most commercialized devices are carbon/carbon systems operated in an organic electrolyte.

However, aqueous systems have many desirable advantages for commercial products, such as their high ionic conductivity, which is useful for optimizing power density [LON 11]. Moreover, there is no risk of thermal runaway when using aqueous electrolytes [GUI 05], which is a key point since ECs are usually operated with high current densities and therefore fast charge/discharge regimes are requested. Cost is also one of the key parameters for justifying the development of supercapacitors in aqueous electrolytes. In this case, the production of cells also has fewer technical drawbacks (no controlled atmosphere for device assembly, no organic solvents, no costly vacuum drying steps, etc.) compared to organic systems, thereby reducing manufacturing costs.

All aqueous electrolyte systems are not limited to an operating voltage of 1.23 V. Several types of secondary batteries exceed this voltage due to the overpotentials for water electrolysis (oxygen/hydrogen). These overpotentials strongly depend on the chemical nature of the electrodes. The best example is with lead-acid batteries that have an operating voltage of about 2 V in concentrated H_2SO_4 acid [LIN 02]. This lead to a whole series of devices using a capacitive carbon negative electrode and a Faradaic positive electrode (battery type electrode), such as PbO_2 or $Ni(OH)_2$ [PEL 04, VAR 95]. These devices have operating voltages of 2.25 and 1.65 V, respectively.

The main result when replacing a capacitive positive electrode with a Faradaic electrode in a supercapacitor design is the drastic increase in the capacitance of the hybrid device due to the very large capacity of the Faradaic electrode compared to that of the capacitive negative carbon electrode: 1,040 C/g for $Ni(OH)_2$ or 810 C/g for PbO_2 versus 250 C/g for an activated carbon electrode (approximately 200 F/g) operated over a 1.23 V voltage range. However, using a Faradaic electrode instead of a capacitive electrode has some drawbacks such as lower cycling ability compared to a standard carbon-based EDLC because of structural and microstructural changes occuring over each cycle, limited power of the battery type electrode, etc. Various solutions have been proposed to overcome these flaws.

As mentioned above, another approach for increasing the energy density of aqueous ECs is the use of pseudocapacitive electrodes such as those based on MnO_2. Despite an increase in cell capacitance, limited voltage in symmetric oxide systems is still an issue for increasing the energy density. In order to tackle this limitation, the negative electrode of symmetrical MnO_2/MnO_2 systems has been replaced, for example, by an activated carbon electrode [HON 02, KHO 06], making it possible to reach greater negative potentials compared to a MnO_2 negative electrode, resulting in a higher cell voltage along with an increase in energy density.

These concepts have been transposed to organic electrolytes by coupling a Faradaic lithium ion intercalation electrode such as graphite in lithium-ion capacitors (also called LiCs) with a capacitive activated carbon electrode, leading to a hybrid device as for the above-mentioned carbon/$Ni(OH)_2$ design.

The manufacturing principles for these different types of hybrid and/or asymmetric devices are detailed in the following sections, with examples illustrating the various designs and main trends for future devices.

4.1. Hybrid devices (asymmetric) in aqueous electrolytes

In the late 1990s, hybrid systems were proposed as an opportunity to increase the limited energy density of conventional carbon/carbon EDLCs [PEL 04, ZHE 03, VAR 95, RAZ 98, VAR 98, LIP 01].

Combining the advantages of long lifetime and high cycling rates of activated carbon negative electrodes, together with those of Faradaic positive electrodes in aqueous electrolytes (high capacity, electrolyte with very good ionic conductivity), these devices were supposed to meet the demands of high energy and power densities. The term "hybrid" or "asymmetric" is used in the literature to refer to these systems.

4.1.1. *Concept and limitations*

The basic principle of a hybrid system is based on the coupling of a battery electrode with an activated carbon electrode, while allowing the latter to operate over the widest possible potential range. Because of the combination with a Faradaic positive electrode operated in a complementary electrochemical potential window, the cell voltage may exceed 1.23 V. In addition, the carbon electrode can be used within its entire electrochemical stability window, while the Faradaic electrode, despite a smaller electrochemical window, has a much greater capacity than that of the capacitive electrode.

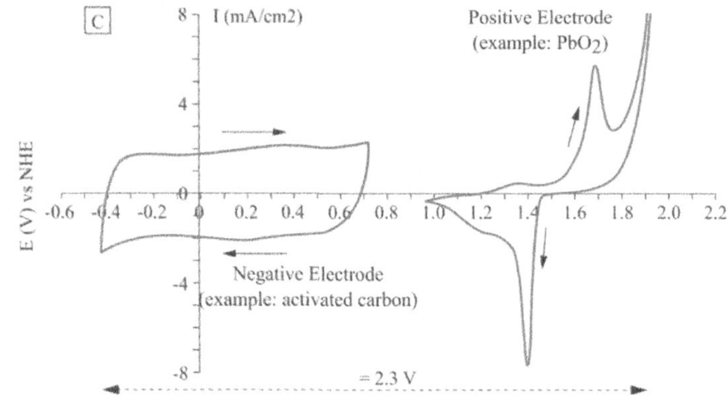

Figure 4.1. *Cyclic voltammogram of a hybrid device: capacitive carbon electrode /aqueous electrolyte(KOH, H_2SO_4)/ Faradaic electrode ($Ni(OH)_2$, PbO_2) [LON 11] . For a color version of the figure, see www.iste.co.uk/brousse/supercapacitors.zip*

It results in a total capacitance greater than that obtained when using symmetrical carbon electrodes. Computations of capacitances expected for such hybrid systems can be found in publications by Pell and Conway [PEL 04] and Zheng [ZHE 03]. An energy density of 50 Wh/kg (of active material) was calculated for a carbon/Ni(OH)$_2$ hybrid system operated over a voltage range of 1.65 V in KOH electrolyte; the same calculation for a carbon/carbon system in H_2SO_4 gives only 7.2 Wh/kg (Table 4.1). Conway obtained the same calculation result for the carbon/Ni(OH)$_2$ system, but the energy density value can reach up to 65 Wh/kg in the case of a carbon/PbO$_2$ system in H_2SO_4 [PEL 04].

Type of cell cathode/ anode	Specific capacity or capacitance	Electrolyte	Operational voltage (V)	The maximum energy density (Wh/kg)	Swing voltage/ operational voltage (%)	Mass distribution cathode/ anode/ electrolyte
Carbon/carbon	280 F/g –280 F/g	5.26 M/L H_2SO_4 in H_2O	1.0	7.16	100	1/1/0.72
Carbon/carbon	120 F/g –120 F/g	0.75 M/L Et_4NBF_4 in PC	3.0	9.41	100	1/1/5.97
$RuO_2.xH_2O$/ $RuO_2.xH_2O$	768 F/g –768 F/g	5.26 M/L H_2SO_4 in H_2O	1.0	26.7	100	1/1
Carbon/ $Li_4Ti_5O_{12}$	120 F/g –168 mA/g	1 M/L $LiPF_6$ in 2 EC:DMC	3.2	34.51	45	3.34/1/7.16
Carbon/WO_2	120 F/g –40 mA/g	1 M/L $LIPF_6$ in 2 EC:DMC	3.6	31.73	38	0.86/1/1.77
$Ni(OH)_2$/carbon	292 mA/g –280 F/g	6.25 M/L KOH in H_2O	1.65	50.35	69	1/3.30/1.97

Table 4.1. *Summary of performances achieved with cells using various electrodes and electrolytes [ZHE 03]*

The technological issue lies in the cyclability of the Faradaic electrode, which is much less than that of the activated carbon capacitive electrode. The energy density values are therefore systematically overestimated because all calculations are made with regard to the full use of the active material of the Faradaic electrode, which is actually difficult to reach for various reasons, the main one being the limited kinetics of the involved redox reactions compared to those of the electrostatic interactions in the carbon electrode. Conway provides the requirements for Faradaic electrodes for application in a hybrid system, including recommendations on the capacitance of a hybrid cell, which should be limited by the capacitive carbon electrode, thereby limiting the depth of discharge (DOD) of the Faradaic electrode [PEL 04]. The Faradaic electrode must therefore be substantially oversized with respect to the device. The charge/discharge regime must also be adapted to the Faradaic electrode rate capability, which is therefore the limiting factor in terms of power density in a hybrid device. This suggests that nanostructured Faradaic electrodes should be preferred to microstructured compounds as they allow a better ionic diffusion in the solid [YU 09, PER 09]. However, since the positive and negative electrodes must have the same face-to-face footprint surface, the faradic electrodes are much thinner than those of carbon, thus allowing higher charge/discharge regimes than with electrodes of conventional batteries.

Today's main hybrid systems are described in the following sections.

4.1.2. *Activated carbon/PbO_2 device*

The half-reactions occurring in lead acid batteries during charging and discharging are described by the double sulfatation mechanism [LIN 02]:

– positive electrode:

$$PbO_2 + 3H^+ + HSO_4^- + 2e^- \leftrightarrow PbSO_4 + 2H_2O \text{ (+1.685 V versus ENH)} \quad [4.2]$$

– negative electrode:

$$Pb + HSO_4^- \leftrightarrow PbSO_4 + H^+ + 2e^- \text{ (–0.356 V versus ENH)} \quad [4.3]$$

In an asymmetric AC/PbO_2 device, the positive electrode and the electrolyte remain the same as in a conventional Pb/PbO_2 battery, but the negative electrode is replaced by a highly reversible carbon electrode (equation [4.4]):

$$nC_6^{x-}(H^+)_x \leftrightarrow nC_6^{(x-2)-}.(H^+)_{x-2} + 2H^+ + 2e^- \text{ (discharge)} \quad [4.4]$$

Stable cycling ability of such a system was observed over more than 10,000 cycles [VOL 02].

However, in the literature, few publications report on hybrid systems [YU 09a, YU 09b, PER 09, VOL 02, CON 03], although some companies do offer products based on this technology. Axion Power International [AXI XX] commercializes a system known as a *PbC®* battery that uses a standard lead-acid battery positive electrode and an activated carbon-based supercapacitor as the negative electrode. The *PbC®* battery is a sealed battery that does not requires any maintenance and therefore has moderate operating costs. Moreover, the manufacturing of *PbC®* batteries is easy as it derives from the production processes of lead-acid batteries.

Apart from the cyclability issue, which is often solved by the addition of an excess of PbO_2 to the positive electrode as a "reserve" of active materials, power is also a problem that has been addressed by various groups [YU 09a, PER 09]. The trend is to use a nanostructured lead oxide electrode (thin films, nanowires, etc.) to maximize the surface area in contact with the electrolyte. Unlike with thin films, PbO_2 nanowires improve the energy density and power density, but the benefit of nanostructuring is lost after a few cycles due to microstructural changes occurring during sulfatation (dissolution/precipitation mechanism).

Alternatively, electrolytes such as methanesulfonic acid, already used in redox flow batteries [HAZ 04], can be implemented in carbon/PbO_2 devices to improve cycling ability [PER 11].

Another strategy to improve the energy density of carbon/PbO_2 devices is to decrease the weight of the electrodes. A significant gain in weight can be achieved by using a carbon-based substrate on which a thick film of Pb or Pb-Sn is deposited [PET 00]. The main bottleneck of this technology is the control of the carbon/lead current collector interface because of corrosion phenomena (formation of $PbSO_4$), which can lead to a complete passivation of the electrode.

Another concept developed by Lam and Louey, at CSIRO Energy Technologies, involved adding a capacitive negative carbon electrode into the Pb-based Faradaic negative electrode to create what they call an "ultra-battery" [LAM 06, LAM 07]. The objective of such an assembly is to increase the power and lifetime of the lead-acid battery through the use of a carbon electrode acting as a "buffer" during charging and discharging; A series of hybrid system has been developped, such as 30 Ah ultrabatteries with 5 h autonomy with over 100,000 charge/discharge cycles [COO 09, FUR 10] showing excellent performances compared to standard VRLA batteries.

4.1.3. *Hybrid AC/$Ni(OH)_2$ device*

Hybrid devices built on activated carbon and nickel oxide or hydroxide, as the negative and positive electrode, respectively, operated in an alkaline electrolyte, were proposed by Russian groups (Elit, ESMA, etc.) at the end of the 1990s [VAR 95, RAZ 98, VAR 98, LIP 01]. These devices are composed of a conventional activated carbon electrode as the negative and a nickel oxide Faradaic positive electrode, similar to those used in Ni-Cd or NiMH batteries. These systems are currently commercialized by various manufacturers. The cell voltage in the charged state is typically 1.5 V, while the capacitance of a cell varies between 3 and 80 kF. A wide variety of designs made from individual cells have been proposed by various research teams [LEC 16, NOH 06, INO 10, PAR 02, NOH 06, WAN 08a].

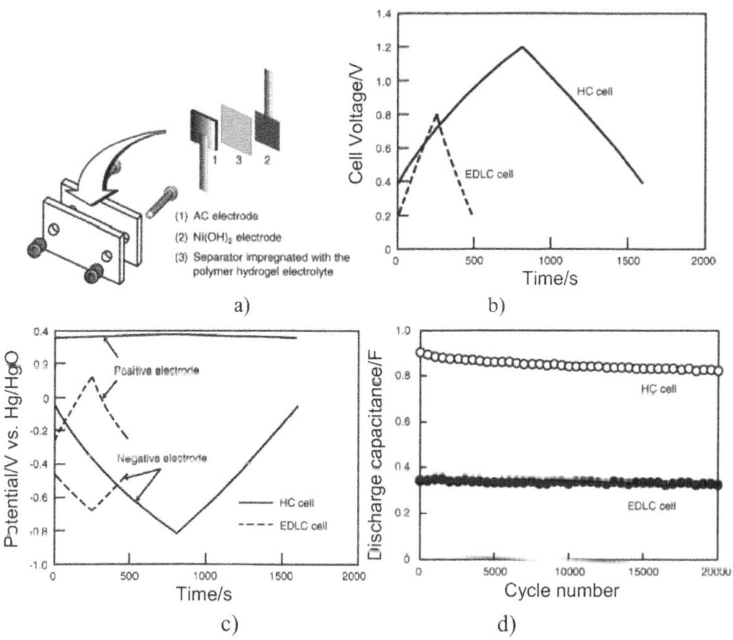

Figure 4.2. *a) Schematic description of the cell; b) comparison between an EDLC in KOH electrolyte and a carbon/$Ni(OH)_2$ hybrid system in a hydrogel based on PAAK and KOH electrolyte, cycled at 1 mA/cm^2; c) details of operational voltage range of each electrode for the two devices; d) the cyclability [NOH 06]*

As predicted by theoretical calculations, the use of a nickel hydroxide electrode improves performance (cell voltage and specific energy) compared with a symmetrical cell based on carbon electrodes. Some information about the basic characteristics and performance of these systems is available in a review article [BUR 00].

Over the past 10 years, the commercialization of this type of cell and increasing interest in supercapacitors have promoted more academic research on this system [BEG 13], with a particular focus on nanostructuring the $Ni(OH)_2$ electrode. With the emergence of new pseudocapacitive electrode materials, original devices have been designed, such as those incorporating a negative VN (pseudocapacitive) electrode and a positive NiO (Faradaic) electrode in 1 M KOH, with a maximum cell voltage of 1.5 V [EUS 13]. Devices with a positive electrode based on conductive polymers and a negative electrode based on activated carbon or conductive polymers have also been proposed [NAO 08, LAF 01a, VIL 03, DIF 01, LAF 03].

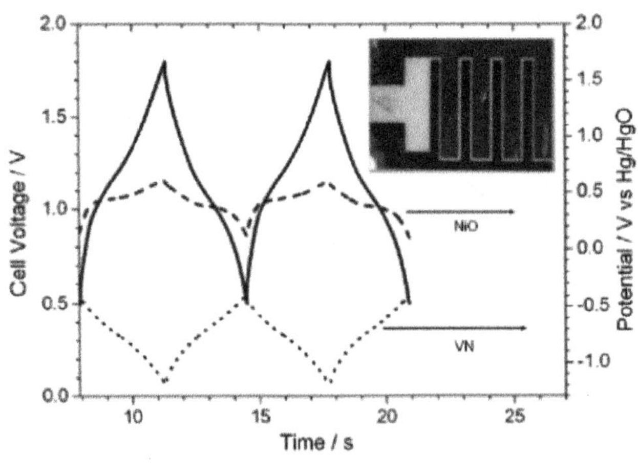

Figure 4.3. *Galvanostatic charge/discharge (1.6 mA cm^{-2}) of a hybrid interdigitated VN(−)//1 M KOH//NiO(+) system. The continuous line is the voltage of the system and the dotted lines represent the electrode potentials. A photo of the interdigitated device is shown in the insert [EUS 13]*

4.2. Asymmetric aqueous supercapacitors

The pseudocapacitive properties of MnO_2 electrodes have given rise to numerous studies focusing on storage mechanism elucidation, capacity improvement, cycling enhancement, structure–capacity relationship, influence of the porous characteristics, etc. [LON 11, BEG 13, BÉL 08]. Only a few authors have argued about the limited electrochemical window of the MnO_2 based electrode in an aqueous electrolyte such as K_2SO_4 [BRO 04, KHO 06, LEE 99, TOU 06]. Some symmetrical MnO_2/MnO_2 devices have been fabricated and characterized [KHO 06, TOU 06]. The advantages associated with the use of MnO_2 (low cost, environmentally friendly, safe and inexpensive manufacturing process, etc.) are balanced by the poor power and energy densities of symmetrical MnO_2/MnO_2 ECs. To overcome these problems, one solution is to use a complementary negative electrode to a positive MnO_2 electrode and thus to design an asymmetric system.

4.2.1. *Activated carbon/MnO_2 devices*

The principle of asymmetric ECs is based on the use of two capacitive or pseudocapacitive electrodes with complementary electrochemical stability windows (Figure 4.4). As for hybrid devices using a Faradaic electrode together with a capacitive electrode, the main purpose is to increase the maximum cell voltage compared to a symmetric cell. The asymmetric AC/MnO_2 system was suggested as a relevant strategy for expanding the cell voltage of the symmetrical MnO_2/MnO_2 EC device. Since first reports by Hong *et al.* and Brousse *et al.* [HON 02, BRO 04], the concept has been widely validated by several teams proving that the strategy is relevant.

Remarkably, the electrochemical signatures of the carbon/MnO_2 system are similar to those of a symmetrical EDLC with two carbon electrodes.

a)

b)

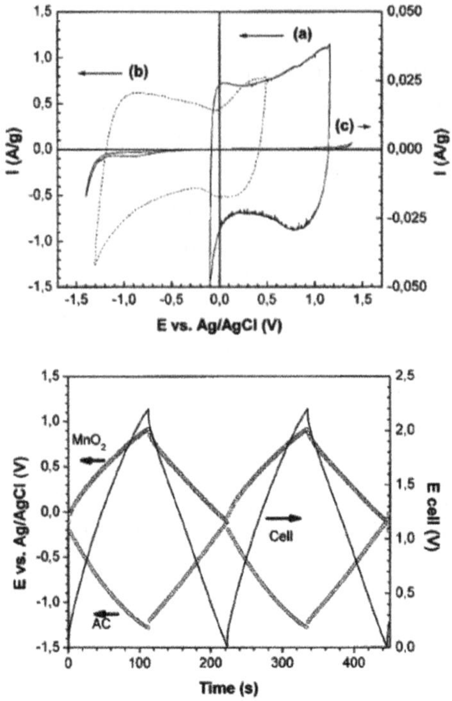

Figure 4.4. *a) Cyclic voltammograms of activated carbon (left) and MnO_2 (right) electrodes showing their complementary electrochemical windows in K_2SO_4 aqueous electrolyte; b) galvanostatic charge/discharge of a complete cell showing the cell voltage and the potentials of the activated carbon and MnO_2 electrodes in a neutral aqueous electrolyte (K_2SO_4, 0.5 M) [BRO 04]*

Table 4.2 summarizes the characteristics of some asymmetric devices based on MnO_2 [BÉL 08]. Most asymmetric AC/MnO_2 ECs have cell voltages of approximately 2.0 V. As a result of this cell voltage increase, activated carbon/MnO_2 combinations provide an energy density greater than 28 Wh·kg^{-1} (normalized to the total mass of active materials), which is almost an order of magnitude greater than the symmetric MnO_2 system, and comparable to conventional symmetrical carbon/carbon EDLCs with non-aqueous electrolytes [LON 11, BÉL 08]. Other materials have been proposed as negative electrodes such as iron oxides (e.g. Fe_3O_4, FeOOH or $LiFeO_2$) [JIN 08, BRO 03, SAN 10] and titanium phosphate [LUO 08, WU 14].

Negative electrode	Positive electrode	Current collector	Electrolyte salt	Cell voltage V	C* $F g^{-1}$	ESR* Ωcm^2	Energy Density* $W h kg^{-1}$	Power Density* $kW kg^{-1}$	Number of cycles	Ref.
AC	MnO_2	Titanium	KCl	2.0	52	—	28.8	0.5	100	22
MnO_2	MnO_2	SS	K_2SO_4	1.0	36	—	3.3	3.08	—	23
Fe_3O_4	MnO_2	SS	K_2SO_4	1.8	21.5	—	8.1	10.2	5,000	23
AC	MnO_2	SS	K_2SO_4	2.2	31	—	17.3	19	10,000	23
AC	MnO_2	Titanium	K_2SO_4	1.5	—	—	7.0	10	23,000	24
MnO_2	MnO_2	Gold	KNO_3	0.6	160	1.56	1.9	3.8	—	25
AC	MnO_2	Gold	KNO_3	2.0	140	0.54	21	123	1,000	25
PANI	MnO_2	Gold	KNO_3	1.2	—	0.57	5.86	42.1	500	26
Ppy	MnO_2	Gold	H_2SO_4	1.4	—	0.52	7.37	62.8	500	26
PEDOT	MnO_2	Gold	KNO_3	1.8	—	0.48	13.5	120.1	500	26
AC	MnO_2	Ni foam	LiOH	1.5	62.4	—	19.5	—	1,500	27
AC	$LiMn_2O_4$	Ni grid	Li_2SO_4	1.8	56	3.3	10.0	2	20,000	28
AC	MnO_2	SS	K_2SO_4	2.0	21	1.3	11.7	—	195,000	29

Abbreviations: activated carbon (AC), polyaniline (PANI), polypyrrole (Ppy), poly(3,4-ethylenedioxythiophene) (PEDOT), stainless steel (SS), capacitance (C), equivalent series resistance (ESR).
**See cited references for further details on experimental conditions and calculations.*

Table 4.2. *Performance of various asymmetric systems presented in the literature using a MnO_2 electrode [BÉL 08]*

The cycling stability of asymmetric AC/MnO_2 ECs has also been studied by various authors [BRO 04, BRO 07, WEI 09] and over 100,000 cycles have been performed with less than a 20 % loss in capacitance. Different parameters appear to influence the lifetime of the manganese dioxide electrode. As such, the observed continuous decrease in capacity of the MnO_2 electrode during cycling is due to volume variations of the oxide particles depending on the initial structure of the selected MnO_2 polymorph [HSI 08]. Another drastic problem is the dissolution of manganese that results in a gradual loss of active material, and hence of capacitance, as the cycles proceed [WEI 09, CHA 09]. This is why it is important to properly balance the masses and capacitances of the positive and negative electrodes to ensure that MnO_2 does not come out of its safe electrochemical stability window during cycling [WAN 06, DEM 11]. One company is currently commercializing an asymmetric supercapacitor based on the principles outlined above [AQU XX].

4.2.2. *Asymmetric carbon/carbon devices*

Recently, asymmetric carbon-based devices [KHO 10, AID 07, DEM 10, RAT 14] have been operated in aqueous electrolytes (Figure 4.5). In these devices, the charging mechanism is not limited to the double layer capacitance, unlike systems in organic electrolytes. An additional Faradaic contribution from oxygen-based moieties (at the carbon surface) plays an important role in aqueous electrolytes. Further ongoing studies aim to evaluate the cycling performance of these particularly promising devices [DEM 10, RAT 14].

4.2.3. *Conclusions on hybrid systems in aqueous electrolytes*

Many other systems based on the same principles have been proposed for asymmetric and hybrid devices in aqueous electrolytes. There are examples of devices using a RuO_2

electrode and an activated carbon electrode [TIA 60]. Another promising way of increasing the capacitance of the electrode is to chemically modify the carbon surface; more specifically, the covalent bonding of a redox active moiety can add a Faradaic contribution to the double-layer capacitance of the activated carbon [PEC 08, POG 11b, KAL 08, SMI 09]. Chemical grafting of organic radicals onto carbon can be carried out using a diazonium approach [TOU 07, ADE 05, BAH 01]. The use of different functionalized surfaces for the positive and negative electrodes leads to an improvement of asymmetric ECs in aqueous electrolytes [ALG 09, POG 12].

a)

b)

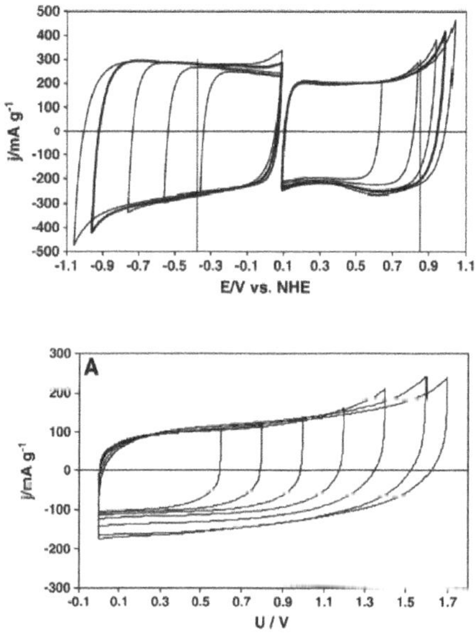

Figure 4.5. *a) Cyclic voltammograms of two activated carbon electrodes in an Na_2SO_4 aqueous electrolyte; b) voltammograms resulting from the assembly of activated carbon electrodes into a cell in the same electrolyte [DEM 10]*

4.3. Hybrid devices in organic electrolytes

Similarly to the strategy that led to the design of hybrid systems in aqueous electrolytes such as carbon/PbO_2 or carbon/$Ni(OH)_2$, hybrid systems can also be designed in organic electrolytes, with a significant increase in energy density since the electrochemical window that defines E_{max} is, by design, larger with organic solvents.

As for hybrid devices in aqueous electrolytes, the Faradaic electrode should:

– increase the total capacitance of the device;

– show a high stability during cycling;

– be able to withstand high charge/discharge regimes.

The design of a hybrid device operated in an organic electrolyte is based on the use of a negative lithium intercalation electrode (graphite, $Li_4Ti_5O_{12}$, TiO_2(B), etc.) and a positive electrode based on activated carbon [BEG 13]. The electrolyte should, therefore, include a lithium salt and conventional electrolytes of Li-ion batteries are usually used (1 M $LiPF_6$ or $LiBF_4$, EC:DEC, EC:DMC, etc.).

4.3.1. *Hybrid titanate/activated carbon systems*

In 2001, Amatucci *et al.* [AMA 01] identified lithium titanate $Li_4Ti_5O_{12}$ as a potentially valuable material due to its high stability during lithium insertion/removal, and its high specific power when synthesized as nanoparticles. The cell voltage of the hybrid device was approximately 3 V.

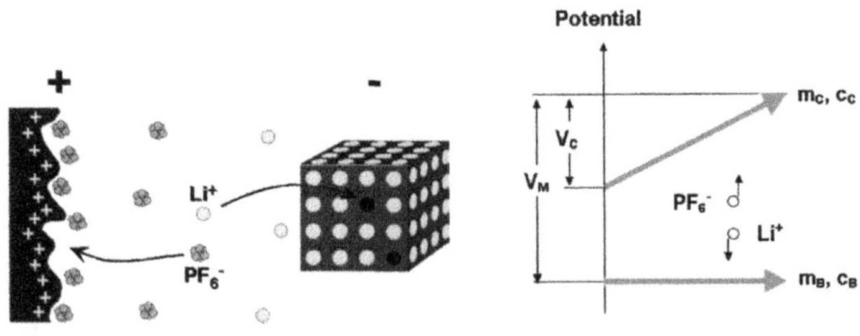

Figure 4.6. *Diagram showing the operation of a hybrid system using a lithium intercalation electrode [ZHE 03, AMA 01]*

The performance obtained was 20 Wh/kg in terms of energy density at a 10 C charge/discharge rate over 4,500 complete charge/discharge cycles.

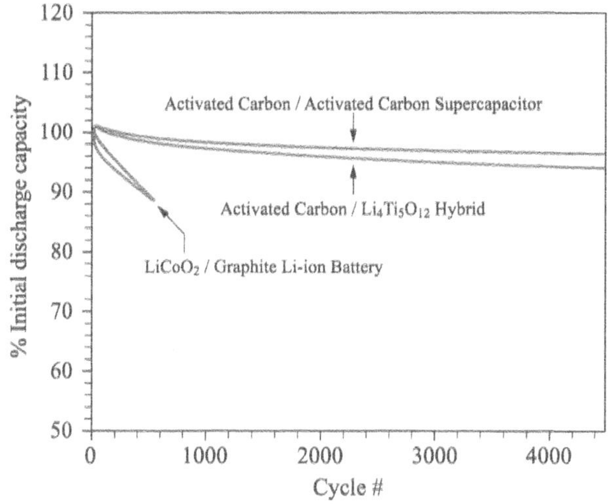

Figure 4.7. *Comparison between the capacitance of an asymmetric hybrid cell, a Li-ion battery and a carbon/carbon supercapacitor [AMA 01]. For a color version of the figure, see www.iste.co.uk/brousse/supercapacitors.zip*

The same team increased the performance by replacing the activated carbon positive electrode with a mix of activated carbon and $LiCoO_2$. The energy density was then 40 Wh/kg and 4 kW/kg over 9,000 charge/discharge cycles [AMA 01].

Research by Amatucci and Du Pasquier has been pursued by various teams (Zaghib [GUE 04], Naoi [NAO 10, NAO 12, NAO 13]) with the aim of improving the performance of the titanate-based negative electrode by synthesizing nanometric materials or nanocomposites with carbon to mitigate the specific power deficit and improve the cyclability. In its device category, this is the most advanced system for which several prototypes have been produced.

Taking advantage of the interesting properties of titanates (high average lithium intercalation potential, good cycling stability, etc.), various research teams have worked on alternative materials such as TiO_2(B) phase [BRO 06]. At ambient temperature, its crystalline structure favors the intercalation of lithium, with a theoretical capacity of 335 mAh/g. Practical capacities are, however, lower than this (200–250 mAh/g), but exceed those of $Li_4Ti_5O_{12}$. Moreover, this oxide can easily be nanostructured (rods, nanowires, nanotubes, nanoribbons, etc.) and exhibits good power performance. Hybrid devices were also made from this material and activated carbon with performances comparable to those reported by Amatucci for the $Li_4Ti_5O_{12}$/carbon system.

4.3.2. *Hybrid graphite/activated carbon systems*

To extend the accessible potential range for the hybrid system, it has been proposed to replace the titanate negative electrode with graphite [SIV 10, KHO 08, WAN 08b, KRA 11b, KIM 11]. The intercalation voltage of lithium is therefore < 0.2 V versus Li^+/Li, which generates a cell voltage up to 4.5 V. Nevertheless, this increase in voltage is balanced by various flaws inherent to the use of graphite: a low intercalation voltage that can lead to safety problems is associated with lithium deposition, moderate cycling ability and poor power response.

It is also needed to pre-lithiate the graphite for a suitable operation (in particular the formation of the SEI on the graphite) of the hybrid device [DEC 12].

The lithiated carbon/activated carbon (porous) system is now the most successful hybrid device. It is commercialized by JMEnergy under the name *Ultimo Cell* (or LiC) [JME 17]. The cyclability is more than 100,000 cycles, but depends on the DOD and the operating temperature range. Moreover, an independent study revealed a 4% decrease in capacitance during the first 4,000 cycles [OMA 12].

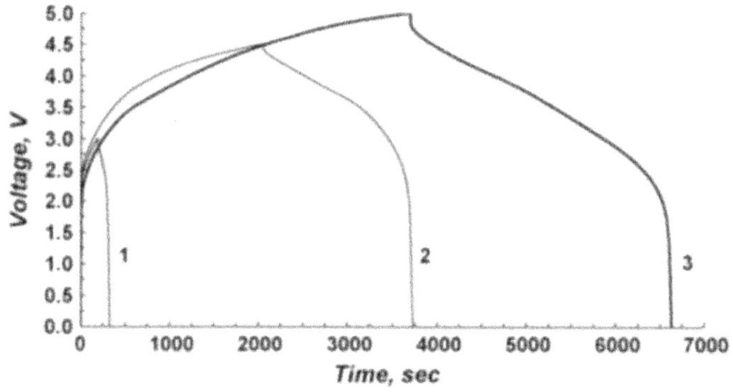

Figure 4.8. *Charge–discharge profile at different voltage limits (1.0–3.0 V; 2.0–4.5 V; 3.0–5.0 V) for a hybrid supercapacitor (LiC) using a negative graphite electrode and a positive activated carbon electrode in a ratio of 1:1 [KHO 08]. For a color version of the figure, see www.iste.co.uk/brousse/super capacitors.zip*

4.4. Conclusion

The use of pseudocapacitive materials makes it possible to achieve high-energy densities, but the increase in power density is one of the main aims of today's research. Research projects are also underway in the field of new materials, their (nano) structuring, shaping and integration into devices, either through the formulation of composite electrodes combining a pseudocapacitive oxide with a meso- or nanostructured carbon, or through the synthesis of materials of high density, or that of metal nitrides. Hybrid or asymmetric devices achieve these high-energy density targets, but still suffer from limitations in terms of power density and cycling ability. In these systems, whether an aqueous or organic electrolyte is selected depends on the cost and the level of safety that is targeted.

Conclusion

Carbon-based supercapacitors have a cycle lifetime which is far greater than that of batteries. They also offer a wider application temperature range (–40 °C to +60 °C) as well as fast and similar charge and recharge rates. This makes them complementary storage devices to batteries in many applications. However, in some cases, they can also be used for electric traction (by replacing batteries), especially when there is no significant weight or volume constraints. In such applications, a fast recharging rate is sought after. This is the case, for example, in tramways and harbor cranes (regenerative braking or potential energy harvesting), in hybrid buses (recharging at station stops for a maximum autonomy of 2 km) and electric boats (island shuttle running for 20 min after a 3 min charge). To increase the energy density of supercapacitors, it is necessary to increase either the capacitance of the materials or the cell voltagel.

Research aimed at increasing the carbon capacitance focus on the following strategies:

– understanding ion transport within the porous structure of carbon;

– developing carbon structures with suitable porosity (nanopores, hierarchical porosity, etc.).

Although the capacitance increase will be limited, fine tuning of the porous carbon structure and texture would allow us to reach 250–300 F/g in organic electrolytes (currently limited to 150 F/g). Another option is to graft active redox moieties onto the carbon surface. In this case, the capacitance increase may be greater, but care must be taken not to compromise cyclability.

Replacing the active material of EDLCs (carbon) with pseudocapacitive materials could lead to a significant increase in the capacitance. Examples are MnO_2 (proton or alkaline cation intercalation/insertion, e.g. Na^+) in an aqueous electrolyte or even Nb_2O_5 in an organic electrolyte. Again, cyclability is still the main issue, together with a possible temperature limitation for a carbonate-based organic medium (Nb_2O_5) or a limited potential window for MnO_2.

Increasing the voltage can be achieved by developing new electrolytes or improved electrolyte formulations. Recent results have shown that ionic liquids such as salts in conventional solvents (e.g. PC) are stable over 3.5 V, but these results remain somewhat questionable due to the technique used to evaluate the cycling stability. Protic ionic liquids have not yet shown a significant improvement in cell capacitance due to their intrinsic limitations (conductivity at $T < 10$ °C, complex drying process, etc.). Another option is to design the carbon electrode – and its porous structure – according to the electrolyte characteristics (e.g., viscosity for instance). In this way, using a mixture of ionic liquids with an activated carbon electrode whose surface is fully accessible (e.g. activated graphene) will enable us to achieve high cell voltage over a wide temperature range. The cost of these electrolytes is still a drawback, but it has been shown that controlling the carbon/electrolyte interface is the key to improving the overall performance.

A research topic recently explored is the development of hybrid (asymmetric) devices that combine a battery type electrode with a supercapacitor electrode (capacitive or pseudocapacitive). An appropriate balancing of the active materials at both electrodes increases the voltage and capacitance of the device. If the energy densities are greatly increased (20 Wh/kg for the LiC, *i.e.* lithiated graphite/porous carbon hybrid in an organic electrolyte), the presence of a faradaic electrode results in limited cyclability and limited power density, known limitations of batteries, unlike EDLCs.

Ultimately, choosing one of the many possible devices depends on its envisioned use.

Bibliography

[ACH 14] ACHOUR A., DUCROS J.-B., PORTO R.-L. *et al.*, "Hierarchical nanocomposite electrodes based on titanium nitride and carbon nanotubes for microsuperca-pacitors", *Nano Energy*, vol. 7, pp. 104–113, 2014.

[ADE 05] ADENIER A., CABET-DELIRY E., CHAUSSE A. *et al.*, "Grafting of nitrophenyl groups on carbon and metallic surfaces without electrochemical induction", *Chemistry of Materials*, vol. 17, pp. 491–501, 2005.

[AID 07] AIDA T., MURAYAMA I., YAMADA K. *et al.*, "High-energy-density hybrid electrochemical capacitor using graphitizable carbon activated with KOH for positive electrode", *Journal of Power Sources*, vol. 166, pp. 462–470, 2007.

[ALG 09] ALGHARAIBEH Z., LIU X., PICKUP P.-G., "An asymmetric anthraquinone-modified carbon/ruthenium oxide supercapacitor", *Journal of Power Sources*, vol. 187, pp. 640–643, 2009.

[AMA 01] AMATUCCI G.-G., BADWAY F., DU PASQUIER A. *et al.*, "An asymmetric hybrid nonaqueous energy storage cell", *Journal of The Electrochemical Society*, vol. 148, pp. A930–A939, 2001.

[AQU 17] AQUION ENERGY, available at www.aquionenergy.com, 2017.

[ARD 90] ARDIZZONE S., FEGONARA G., TRASATTI S., "Inner and outer active surface of RuO_2 electrodes", *Electrochimica Acta*, vol. 35, pp. 263–267, 1990.

[AUG 13] AUGUSTYN V., COME J., LOWE M.-A. *et al.*, "High-rate electrochemical energy storage through Li^+ intercalation pseudo capacitance", *Nature Materials*, vol. 12, pp. 518–522, 2013.

[AUG 14] AUGUSTYN V., SIMON P., DUNN B., "Pseudocapacitive oxide materials for high-rate electrochemical energy storage", *Energy & Environmental Science*, vol. 7, pp. 1597–1614, 2014.

[AXI 17] AXION POWER, available at www.axionpower.com, 2017.

[BAH 01] BAHR J.-L., TOUR J.-M., "Highly functionalized carbon nanotubes using *in situ* generated diazonium compounds", *Chemistry of Materials*, vol. 13, pp. 3823–3824, 2001.

[BÉG 13] BÉGUIN F., FRACKOWIAK E., LU M. (eds.), *Supercapacitors: Materials, Systems and Applications*, Wiley, London, 2013.

[BÉG 14] BÉGUIN F., PRESSER V., BALDUCCI A. *et al.*, "Carbons and electrolytes for advanced supercapacitors", *Advanced Materials*, vol. 14, pp. 2219–2251, 2014.

[BÉL 08] BÉLANGER D., BROUSSE T., LONG J.-W., "Manganese oxides: battery materials make the leap to electrochemical capacitors", *Interface*, vol. 17, pp. 49–52, 2008.

[BÉL 11] BÉLANGER D., PINSON J., "Electrografting: a powerful method for surface modification", *Chemical Society Reviews*, vol. 40, pp. 3995–4048, 2011.

[BIC 10] BICHAT M.-P., RAYMUNDO-PIÑERO E., BÉGUIN F., "High voltage super capacitor built with seaweed carbons in neutral aqueous electrolyte", *Carbon*, vol. 48, pp. 4351–4361, 2010.

[BOI 13] BOISSET A., MENNE S., JACQUEMIN J., "Deep eutectic solvents based on *N*-methylacetamide and a lithium salt as suitable electrolytes for lithium-ion batteries", *Physical Chemistry Chemical Physics*, vol. 15, pp. 20054–20063, 2013.

[BOU 13] BOUHTIYYA S., LUCIO PORTO R., LAÏK B. *et al.*, "Application of sputtered ruthenium nitride thin films as electrode material for energy-storage devices", *Scripta Materialia*, vol. 68, pp. 659–662, 2013.

[BRA 12a] BRANDT A., ISKEN P., LEX-BALDUCCI A. *et al.*, "Adiponitrile-based electrochemical double layer capacitor", *Journal of Power Sources*, vol. 204, pp. 213–219, 2012.

[BRA 12b] BRANDT A., BALDUCCI A., "The Influence of pore structure and surface groups on the performance of high voltage electrochemical double layer capacitors containing adiponitrile-based electrolyte", *Journal of the Electrochemical Society*, vol. 159, pp. A2053–A2059, 2012.

[BRA 13] BRANDT A., PIRES J., ANOUTI M., *et al.*, "An investigation about the cycling stability of supercapacitors containing protic ionic liquids as electrolyte components", *Electrochimica Acta*, vol. 108, pp. 226–231, 2013.

[BRA 14] BRACHET M., BROUSSE T., LE BIDEAU J., "All solid-state symmetrical activated carbon electrochemical double layer capacitors designed with ionogel electrolyte", *ECS Electrochemistry Letters*, vol. 3, no. 11, pp. A112–A115, 2014.

[BRA 16] BRACHET M., GABORIAU D., GENTILE G. *et al.*, "Solder-reflow resistant solid-state micro super capacitors based on ionogels", *Journal of Materials Chemistry A*, vol. 4, pp. 11835–11843, 2016.

[BRO 03] BROUSSE T., BÉLANGER D., "A hybrid Fe_3O_4 MnO_2 capacitor in mild aqueous electrolyte", *Electrochemical and Solid State Letters*, vol. 6, pp. A244–A248, 2003.

[BRO 04] BROUSSE T., TOUPIN M., BÉLANGER D., "A hybrid activated carbon-manganese dioxide capacitor using a mild aqueous electrolyte", *Journal of Electrochemical Society*, vol. 151, pp. A614–A622, 2004.

[BRO 06] BROUSSE T., MARCHAND R., TABERNA P.-L. *et al.*, "TiO_2 (B)/activated carbon non-aqueous hybrid system for energy storage", *Journal of Power Sources*, vol. 158, pp. 571–577, 2006.

[BRO 07] BROUSSE T., TABERNA P.-L., CROSNIER O. *et al.*, "Long term cycling behavior of asymmetric activated carbon/MnO_2 aqueous electrochemical capacitor", *Journal of Power Sources*, vol. 173, pp. 633–641, 2007.

[BRO 13] BROUSSE T., BEGUIN F., "Les superconducteurs", in MOSSERI R., JEANDEL C., (eds.) *L'énergie à découvert*, CNRS éditions, Paris, 2013.

[BRO 15] BROUSSE T., BÉLANGER D., LONG J.W., "To be or not to be pseudocapacitive?", *Journal of the Electrochemical Society*, vol. 162, no. 5, pp. A5185–A5189, 2015.

[BUH 07] BUHA J., DJERDJ I., ANTONIETTI M. *et al.*, "Thermal transfor-mation of metal oxide nanoparticles into nanocrystalline metal nitrides using cyanamide and urea as nitrogen source", *Chemistry of Materials*, vol. 19, pp. 3499–3505, 2007.

[BUR 00] BURKE A., "Ultracapacitors: why, how, and where is the technology", *Journal of Power Sources*, vol. 91, pp. 37–50, 2000.

[CHA 09] CHANG J.-K., HUANG C.-H., LEE M.-T. *et al.*, "Physicochemical factors that affect the pseudocapacitance and cyclic stability of Mn oxide electrodes", *Electrochimica Acta*, vol. 54, no. 12, pp. 3278–3284, 2009.

[CHE 10] CHEN S., ZHU J., WU X. *et al.*, "Graphene oxide-MnO_2 nanocomposites for supercapacitors", *ACS Nano*, vol. 4, no. 5, pp. 2822–2830, 2010.

[CHE 11] CHENG F., HE C., SHUA D. *et al.*, "Soft-template synthesis of vanadium oxynitride-carbon nanomaterials for supercapacitors", *Materials Chemistry and Physics*, vol. 131, pp. 268–273, 2011.

[CHM 06] CHMIOLA J., YUSHIN G., GOGOTSI Y. *et al.*, "Anomalous increase in carbon capacitance at pore size below 1 nm", *Science*, vol. 313, pp. 1760–1763, 2006.

[CHO 05] CHOI D., KUMTA P.-N., "Chemically synthesized nanostructured VN for pseudocapacitor application", *Electrochemical and Solid-State Letters*, vol. 8, no. 8, pp. A418–A422, 2005.

[CHO 06] CHOI D., GEORGE E., BLOMGREN E. *et al.*, "Fast and reversible surface redox reaction in nanocrystalline vanadium nitride supercapacitors", *Advanced Materials*, vol. 18, pp. 1178–1182, 2006.

[CHO 09] CHOUDHURY N.-A., SAMPATH S., SHUKLA A.-K., "Hydrogel-polymer electrolytes for electrochemical capacitors: an overview", *Energy and Environmental Science*, vol. 2, pp. 55–67, 2009.

[CHO 11] CHOI W., HARADA D., OYAIZU K. *et al.*, "Aqueous electrochemistry of poly(vinylanthraquinone) for anode-active materials in high-density and rechargeable polymer/air batteries", *Journal of the American Chemical Society*, vol. 133, pp. 19839–19843, 2011.

[COM 09] COMBELLAS C., JIANG D., PINSON J. *et al.*, "Steric effects in the reaction of aryl radicals on surfaces", *Langmuir*, vol. 25, pp. 286–293, 2009.

[CON 91] CONWAY B.-E.,"Transition from supercapacitor to battery behavior in electrochemical energy-storage", *Journal of The Electrochemical Society*, vol. 138, no. 6, pp. 1539–1548, 1991.

[CON 98] CONWAY B.-E., PELL W.-G., LIU T.-C. *et al.*, "Behavior of molybdenum nitrides as materials for electrochemical capacitors, comparison with ruthenium oxide", *Journal of the Electrochemical Society*, vol. 145, no. 6, pp. 1882–1888, 1998.

[CON 99] CONWAY B.-E., *Electrochemical Supercapacitors: Scientific Fundamentals and Technological Applications*, Kluwer Academic/ Plenum Publishers, New York, 1999.

[CON 03] CONWAY B.-E., PELL W.-G., "Double-layer and pseudocapacitance types of electrochemical capacitors and their applications to the development of hybrid devices", *Journal of Solid State Electrochemistry*, vol. 7, pp. 637–644, 2003.

[CON 13] CONWAY B.-E., *Electrochemical Supercapacitors, Scientific Fundamentals and Technological Applications*, Kluwer Academic, Dordrecht, 2013.

[COO 09] COOPER A., FURAKAWA J., LAM L., *et al.*, "The UltraBattery–a new battery design for a new beginning in hybrid electric vehicle energy storage", *Journal of Power Sources*, vol. 188, pp. 642–649, 2009.

[DEC 12] DECAUX C., LOTA G., RAYMUNDO-PIÑERO E. *et al.*, "Electrochemical performance of a hybrid lithium-ion capacitor with a graphite anode preloaded from lithium bis(trifluoromethane) sulfonimide-based electrolyte", *Electrochimica Acta*, vol. 86, pp. 282–286, 2012.

[DEM 10] DEMARCONNAY L., RAYMUNDO-PINERO E., BÉGUIN F., "A symmetric carbon/carbon supercapacitor operating at 1.6 V by using a neutral aqueous solution", *Electrochemistry Communications*, vol. 12, pp. 1275–1278, 2010.

[DEM 11] DEMARCONNAY L., RAYMUNDO-PINERO E., BÉGUIN F., "Adjustment of electrodes potential window in an asymmetric carbon/MnO_2 supercapacitor", *Journal of Power Sources*, vol. 196, no. 1, pp. 580–586, 2011.

[DIF 01] DI FABIO A., GIORGI A., MASTRAGOSTINO M. *et al.*, "Carbon-poly (3-methylthiophene) hybrid supercapacitors", *Journal of Electrochemical Society*, vol. 148, pp. A845–A850, 2001.

[DON 11] DONG S., CHEN X., GU L. *et al.*, "TiN-VN composites with core shell structure for supercapacitors", *Materials Research Bulletin*, vol. 46, pp. 835–839, 2011.

[DUP 04] DU PASQUIER A., PLITZ I., GURAL J. *et al.*, "Power-ion battery: bridging the gap between Li-ion and supercapacitor chemistries", *Journal of Power Sources*, vol. 136, pp. 160–170, 2004.

[EUS 13] EUSTACHE E., FRAPPIER R., LUCIO PORTO R. *et al.*, "Asymmetric electrochemical capacitor microdevice designed with vanadium nitride and nickel oxide thin film electrodes", *Electrochemistry Communications*, vol. 28, pp. 104–106, 2013.

[FED 08] FEDOROV M.-V., KORNYSHEV A.-A., "Ionic liquid near a charged wall: structure and capacitance of electrical double layer", *Journal of Physical Chemistry B.*, vol. 112, pp. 11868–11872, 2008.

[FUR 10] FURUKAWA J., TAKADA T., MONMA D. *et al.*, "Further demonstration of the VRLA-type ultrabattery under medium-HEV duty and development of the flooded-type ultrabattery for micro-HEV applications", *Journal of Power Sources*, vol. 195, pp. 1241–1245, 2010.

[GAD 08] GADIOU R., DIDION A., GEARBA R.-I. *et al.*, "Synthesis and properties of new nitrogen-doped nanostructured carbon materials obtained by templating of mesoporous silicas with aminosugars", *Journal of Physics and Chemistry of Solids*, vol. 69, pp. 1808–1814, 2008.

[GAO 12a] GAO Q., DEMARCONNAY L., RAYMUNDO-PIÑERO E. *et al.*, "Exploring the large voltage range of carbon/carbon supercapacitors in aqueous lithium sulfate electrolyte", *Energy Environmental Science*, vol. 5, no. 9611–9617, 2012.

[GAO 12b] GAO Z., ZHANG H., CAO G. *et al.*, "Template synthesis of mesoporous nanocrystalline VN electrode materials and its supercapacitive behavior", *Journal of Inorganic Materials*, vol. 27, pp. 1261–1265, 2012.

[GAO 13a] GAO H., TING Y.-J., KHERANI N.-P. *et al.*, "Ultra-high-rate all-solid pseudo-capacitive electrochemical capacitors", *Journal of Power Sources*, vol. 222, pp. 301–304, 2013.

[GAO 13b] GAO P.-C., SIMON P., FAVIER F., "Silicon carbide with tunable ordered mesoporosity", *Microporous and Mesoporous Materials*, vol. 180, pp. 172–177, 2013.

[GAO 13c] GAO Z., ZHANG H., CAO G. *et al.*, "Spherical porous VN and NiOx as electrode materials for asymmetric supercapacitor", *Electrochimica Acta*, vol. 87, pp. 375–380, 2013.

[GAO 14] GAO P.-C., RUSSO P.-A., CONTE D.-E. *et al.*, "Morphology effects on the supercapacitive electrochemical performances of iron oxide/reduced graphene oxide nanocomposites", *ChemElectroChem*, vol. 1, no. 4, pp. 747–754, 2014.

[GHI 11] GHIMBEU C.-M., RAYMUNDO-PIÑERO E., FIOUX PP., "Vanadium nitride/nanocomposites as electrodes for supercapacitors", *Journal of Materials Chemistry*, vol. 21, pp. 13268–13275, 2011.

[GHO 09] GHODBANE O., PASCAL J.-L., FAVIER F., "Microstructural effects on charge-storage properties in MnO_2-based electrochemical supercapacitors", *ACS Applied Materials & Interfaces*, vol. 1, no. 5, pp. 1130–1139, 2009.

[GHO 12] GHODBANE O., ATAHERIAN F., WU N.-L. *et al.*, "*In situ* crystallographic investigations of charge storage mechanisms in MnO_2-based electrochemical capacitors", *Journal of Power Sources*, vol. 206, pp. 454–262, 2012.

[GHO 13] GHODBANE O., LOURO M., COUSTAN L. *et al.*, "Microstructural and morphological effects on charge storage properties in MnO_2-Carbon Nanofibers based supercapacitors", *Journal of the Electrochemical Society*, vol. 160, pp. A2315–A2321, 2013.

[GIO 09] GIORDANO C., ERPEN C., YAO W. *et al.*, "Metal nitride and metal carbide nanoparticles by a soft urea pathway", *Chemistry of Materials*, vol. 21, pp. 5136–5144, 2009.

[GLU 10] GLUSHENKOV A.-M., HULICOVA-JURCAKOVA D., LLEWELLYN D. *et al.*, "Structure and capacitive properties of porous nanocrystalline VN prepared by temperature-programmed ammonia reduction of V_2O_5", *Chemistry of Materials*, vol. 22, pp. 914–921, 2010.

[GOU 15] GOUBARD-BRETESCHÉ N., CROSNIER O., PAYEN C. *et al.*, "Nanocrystalline $FeWO_4$ as a pseudocapacitive electrode material for high volumetric energy density supercapacitors operated in an aqueous electrolyte", *Electrochemistry Communications*, vol. 57, pp. 61–64, 2015.

[GOU 16] GOUBARD-BRETESCHÉ N., CROSNIER O., FAVIER F. *et al.*, "Improving the volumetric energy density of supercapacitors", *Electrochimica Acta*, vol. 206, pp. 458–463, 2016.

[GUE 04] GUERFI A., CHAREST P., KINOSHITA K. *et al.*, "Nano electronically conductive titanium-spinel as lithium ion storage negative electrode", *Journal of Power Sources*, vol. 126, pp. 163–168, 2004.

[GUI 05] GUILLEMET P., DUGAS R., SCUDELLER Y. *et al.*, "Electro-thermal analysis of a hybrid activated carbon/MnO_2 aqueous electrochemical capacitor", *207th Meeting of the ElectroChemical Society*, Quebec, Canada, May 2005.

[HAZ 04] HAZZA A., PLETCHER D., WILLS R., "A novel flow battery: a lead acid battery based on an electrolyte with soluble lead(II): Part I. Preliminary studies", *Physical Chemistry Chemical Physics*, vol. 6, pp. 1773–1778, 2004.

[HEC 98] HECTOR A.-L., HENSHAW G., KOMAROV A.-V., *et al.*, "Nitrides from solid state metathesis reactions: synthesis and mechanistics", *Journal of Materials Processing Technology*, vol. 77, pp. 103–107, 1998.

[HON 02] HONG M.-S., LEE S.-H., KIM S.-W., "Use of KCl aqueous electrolyte for 2 V manganese oxide/activated carbon hybrid capacitor", *Electrochemical and Solid-State Letters*, vol. 5, pp. A227–230, 2002.

[HSI 08] HSIEH Y.-C., LEE K.-T., LIN Y.-P., *et al.*, "Investigation on capacity fading of aqueous $MnO_2 \cdot nH_2O$ electrochemical capacitor", *Journal of Power Sources*, vol. 177, pp. 660–664, 2008.

[HUL 05] HULICOVA D., YAMASHITA J., SONEDA Y. *et al.*, "Supercapacitors prepared from melamine-based carbon", *Chemistry of Materials*, vol. 17, pp. 1241–1247, 2005.

[HUL 09] HULICOVA-JURCAKOVA D., PUZIY A.-M., PODDUBNAYA O.-I. *et al.* "Highly stable performance of supercapacitors from phosphorus-enriched carbons", *Journal of the American Chemical*, vol. 131, pp. 5026–5027, 2009.

[INO 10] INOUE H., NAMBA Y., HIGUCHI E., "Preparation and characterization of Ni-based positive electrodes for use in aqueous electrochemical capacitors", *Journal of Power Sources*, vol. 195, pp. 6239–6244, 2010.

[JIN 08] JIN W.-H., CAO G.-T., SUN J.-Y., "Hybrid supercapacitor based on MnO_2 and columned FeOOH using Li_2SO_4 electrolyte solution", *Journal of Power Sources*, vol. 175, pp. 686–691, 2008.

[JME 17] JM ENERGY, available at http://www.jmenergy.co.jp/en/index.html, 2017.

[KAL 08] KALINATHAN K., DESROCHES D.-P., LIU X. *et al.*, "Anthraquinone modified carbon fabric supercapacitors with improved energy and power densities", *Journal of Power Sources*, vol. 181, pp. 182–185, 2008.

[KAR 12] KARTACHOVA O., GLUSHENKOV A.-M., CHEN Y. *et al.*, "Electrochemical capacitance of mesoporous tungsten oxynitride in aqueous electrolytes", *Journal of Power Sources*, vol. 220, pp. 298–305, 2012.

[KHO 05] KHOMENKO V., RAYMUNDO-PINERO E., BEGUIN F., "Optimisation of an asymmetric manganese oxide/activated carbon capacitor working at 2 V in aqueous medium", *Journal of Power Sources*, vol. 153, pp. 183–190, 2005.

[KHO 06] KHOMENKO V., RAYMUNDO-PIÑERO E., FRACKOWIAK E., *et al.*, "Optimisation of an asymmetric manganese oxide/activated carbon capacitor working at 2 V in aqueous medium, High-voltage asymmetric supercapacitors operating in aqueous electrolyte", *Applied Physics A-Materials Science and Processing*, vol. 82, pp. 567–573, 2006.

[KHO 08] KHOMENKO V., RAYMUNDO-PINERO E., BÉGUIN F., "High-energy density graphite/AC capacitor in organic electrolyte", *Journal of Power Sources*, vol. 177, pp. 643–651, 2008.

[KHO 10] KHOMENKO V., RAYMUNDO-PINERO E., BÉGUIN F., "A new type of high energy asymmetric capacitor with nanoporous carbon electrodes in aqueous electrolyte", *Journal of Power Sources*, vol. 195, pp. 4234–4241, 2010.

[KIM 11] KIM J.-H., KIM J.-S., LIM Y.-G. *et al.*, "Effect of carbon types on the electrochemical properties of negative electrodes for Li-ion capacitors", *Journal of Power Sources*, vol. 196, pp. 10490–10495, 2011.

[KON 11a] KONDRAT S., GEORGI N., FEDOROV M.-V. *et al.*, "A superionic state in nano-porous double-layer capacitors: insights from Monte Carlo simulations", *Physical Chemistry Chemical Physics*, vol. 13, pp. 11359–11366, 2011.

[KON 11b] KONDRAT S., KORNYSHEV A.-A., "Superionic state in double-layer capacitors with nanoporous electrodes", *Journal of Physics: Condensed Matter*, vol. 23, no. 022201, 2011.

[KON 14] KONDRAT S., WU P., QIAO R., *et al.*, "Accelerating charging dynamics in subnanometre pores", *Nature Materials*, vol. 13, pp 387–393, 2014.

[KRA 11a] KRAUSE A., BALDUCCI A., "High voltage electrochemical double layer capacitor containing mixtures of ionic liquids and organic carbonate as electrolytes", *Electrochemistry Communications*, vol. 13, pp. 814–817, 2011.

[KRA 11b] KRAUSE A., KOSSYREV P., OLJACA M. *et al.*, "Electrochemical double layer capacitor and lithium-ion capacitor based on carbon black", *Journal of Power Sources*, vol. 196, pp. 8836–8842, 2011.

[KUB 13] KUBO S., WHITE R.-J., TAUER K. *et al.*, "Flexible coral-like carbon nanoarchitectures via a dual block copolymer–latex templating approach", *Chemistry of Materials*, vol. 25, pp. 4781–4790, 2013.

[LAF 01a] LAFORGUE A., SIMON P., FAUVARQUES J.-F. *et al.*, "*Hybrid supercapacitors based on activated carbons and conducting polymers*", *Journal of The Electrochemical Society*, vol. 148, no. 10, pp. A1130–A1134, 2001.

[LAF 01b] LAFORGUE A., SIMON P., FAUVARQUES J.-F., "Chemical synthesis and characterization of fluorinated polyphenylthiophenes: application to energy storage", *Synthetic Metals*, vol. 123, pp. 311–319, 2001.

[LAF 03] LAFORGUE A., SIMON P., FAUVARQUE J.-F. *et al.* "Activated carbon/conducting polymer hybrid supercapacitors", *Journal of Electrochemical Society*, vol. 150, pp. A645–651, 2003.

[LAH 11] LAHEÄÄR A., JÄNES A., LUST E., "Electrochemical properties of carbide-derived carbon electrodes in non-aqueous electrolytes based on different Li-salts", *Electrochimica Acta*, vol. 56, pp. 9048–9055, 2011.

[LAH 12] LAHEÄÄR A., JÄNES A., LUST E., "Lithium bis(oxalato)borate as an electrolyte for micromesoporous carbide-derived carbon based supercapacitors", *Journal of Electroanalytical Chemistry*, vol. 669, pp. 67–72, 2012.

[LAM 06] LAM L.-T., LOUEY R., "Development of ultra-battery for hybrid-electric vehicle applications", *Journal of Power Sources*, vol. 158, pp. 1140–1148, 2006.

[LAM 07] LAM L.-T., LOUEY R., HAIGH N.-P. *et al.* "VRLA, ultrabattery for high-rate partial-state-of-charge operation", *Journal of Power Sources*, vol. 174, pp. 16–29, 2007.

[LAR 08] LARGEOT C., PORTET C., CHMIOLA J. *et al.*, "Relation between the ion size and pore size for an electric double-layer capacitor", *Journal of the American Chemical Society*, vol. 130, no. 9, pp. 2730–2731, 2008.

[LEB 11a] LE BIDEAU J., VIAU L., VIOUX A., "Ionogels, ionic liquid based hybrid materials", *Chemical Society Reviews*, vol. 40, pp. 907–925, 2011.

[LEB 1b] LEBÈGUE E., MADEC L., BROUSSE T. *et al*, "Modification of activated carbons based on diazonium ions *in situ* produced from aminobenzene organic acid without addition of other acid", *Journal of Materials Chemistry*, vol. 21, no. 12221–12223, 2011.

[LEB 14] LEBÈGUE E., BROUSSE T., GAUBICHER J., *et al.*, "Toward fully organic rechargeable charge storage devices based on carbon electrodes grafted with redox molecules", *Journal of Materials Chemistry A*, vol. 2, pp. 8599–8602, 2014.

[LEC 13] LE COMTE A., POGNON G., BROUSSE T., *et al.*, "Determination of the quinone-loading of a modified carbon powder-based electrode for electrochemical capacitor", *Electrochemistry*, vol. 81, no. 10, pp. 863–866, 2013.

[LEC 14] LE COMTE A., BROUSSE T., BÉLANGER D., "Simpler and greener grafting method for improving the stability of anthraquinone-modified carbon electrode in alkaline media", *Electrochimica Acta*, vol. 137, pp. 447–453, 2014.

[LEC 15] LE COMTE A., CHHIN D., GAGNON A. *et al.*, "Spontaneous grafting of 9,10-phenanthrenequinone on porous carbon as active electrode material in electrochemical capacitor in alkaline electrolyte", *Journal of Materials Chemistry A*, vol. 3, pp. 6146, 2015.

[LEC 16] LE COMTE A., BROUSSE T., BÉLANGER D., "New generation of hybrid carbon/$Ni(OH)_2$ electrochemical capacitor using functionalized carbon electrode", *Journal of Power Sources*, vol. 326, pp. 702–710, 2016.

[LEJ 13] LE JOURNAL DES ENTREPRISES, available at www.lejournaldesentreprises.com/regionale/Bretagne/lorient-un-premier-bateau-a-passagers-sans-emission-de-co2-pour-stx-france-18-09-2013-204136.php, 2013.

[LEE 99] LEE H.-Y., GOODENOUGH J.-B., "Supercapacitor behavior with KCl electrolyte", *Journal of Solid State Chemistry*, vol. 144, pp. 220–223, 1999.

[LIN 02] LINDEN D., REDDY T.-B., *Handbook of Batteries*, 3rd ed., McGraw-Hill, New York, 2002.

[LIN 11] LIN R., TABERNA P.-L., FANTINI S. *et al.*, "Capacitive energy storage from –50 °C to 100 °C using an ionic liquid electrolyte", *Journal of Physical Chemistry Letters*, vol. 19, pp. 2396–2401, 2011.

[LIN 15] LIN T.., CHEN I.W., LIU F.X. *et al.*, "Nitrogen-doped mesoporous carbon of extraordinary capacitance for electrochemical energy storage" *Science*, vol. 350, no. 6267, pp. 1508–1513, 2015.

[LIP 01] LIPKA S.-M., REISNER D.-E., DAI J.*et al.*, "Asymmetric-type electrochemical supercapacitor development under the atp – an update", *Proceedings of the 11th International Seminar on Double Layer Capacitors*, Deerfield Beach, 2001.

[LON 11] LONG J.W., BÉLANGER D., BROUSSE T. *et al.*, "Asymmetric electrochemical capacitors; stretching the limits of aqueous electrolytes", *MRS Bulletin*, vol. 36, no. 7, pp. 513–522, 2011.

[LU 11] LU Q., ZHOU Y., "Synthesis of mesoporous polythiophene/MnO_2 nanocomposite and its enhanced pseudocapacitive properties", *Journal of Power Sources*, vol. 196, no. 8, pp. 4088–4094, 2011.

[LU 13] LU X., YU M., ZHAI T. *et al.*, "High energy density asymmetric quasi-solid-state supercapacitor based on porous vanadium nitride nanowire anode", *Nano Letters*, vol. 13, pp. 2628–2633, 2013.

[LUC 12] LUCIO PORTO R., FRAPPIER R., DUCROS J.-B. *et al.*, "Titanium and vanadium oxynitride powders as pseudocapacitive materials for electrochemical capacitors", *Electrochimica Acta*, vol. 82, pp. 257–262, 2012.

[LUO 08] LUO J.-Y., LIU J.-L., HE P. *et al.*, "A novel $LiTi_2(PO_4)_3/MnO_2$ hybrid supercapacitor in lithium sulfate aqueous electrolyte", *Electrochimica Acta*, vol. 53, pp. 8128–8133, 2008.

[MA 13] MA T.-Y., LIU L., YUAN Z.-Y., "Direct synthesis of ordered mesoporous carbons", *Chemical Society reviews*, vol. 42, pp. 3977–4003, 2013.

[MEN 05] MENG Y., GU D., ZHANG F. *et al.*, "Ordered mesoporous polymers and homologous carbon frameworks : amphiphilic surfactant templating and direct transformation", *Angewandte Chemie International Edition*, vol. 117, pp. 7215–7221, 2005.

[MEN 13] MENNE S., PIRES J., ANOUTI M. *et al.*, "Protic ionic liquids as electrolytes for lithium-ion batteries", *Electrochemistry Communications*, vol. 31, pp. 39–41, 2013.

[MER 12] MERLET C., ROTENBERG B., MADDEN P.-A. *et al* "On the molecular origin of supercapacitance in nanoporous carbon electrodes", *Nature Materials*, vol. 11, pp. 306–310, 2012.

[MER 13] MERLET C., PEAN C., ROTENBERG B. *et al.*, "Hidden ions store charge more efficiently in supercapacitors", *Nature Communications*, vol. 4, no. 2701, DOI 10.1038/ncomms3701, 2013.

[MIL 08] MILLER J.-R., BURKE A.-F., "Electrochemical capacitors: challenges and opportunities for real world applications", *The Electrochemical Society Interface*, vol. 1, pp. 52–56, 2008.

[MOU 16] MOURAD E., COUSTAN L., LANNELONGUE P. *et al.*, "Biredox ionic liquids with solid-like redox density in the liquid state for high-energy supercapacitors", *Nature Material,* doi:10.1038/nmat 4808, published online, 2016.

[MYS 12] MYSYK R., GAO Q., RAYMUNDO-PIÑERO E., *et al.*, "Microporous carbons finely-tuned by cyclic high-pressure low-temperature oxidation and their use in electrochemical capacitors", *Carbon*, vol. 50, pp. 3367–3374, 2012.

[NAO 08] NAOI K., MORITA M., "Advanced polymers as active materials and electrolytes for ECs and hybrid capacitor systems", *Electrochemical Society Interface*, vol. 8, pp. 44–48, 2008.

[NAO 10] NAOI K., ISHIMOTO S., ISOBE Y., *et al.*, "High-rate nano-crystalline Li4Ti5O12 attached on carbon nano-fibers for hybrid supercapacitors", *Journal of Power Sources*, vol. 195, no. 18, pp. 6250–6254, 2010.

[NAO 12] NAOI K., ISHIMOTO S., MIYAMOTO J.-I. *et al.*, "Second generation 'nanohybrid supercapacitor': evolution of capacitive energy storage devices", *Energy & Environmental Science*, vol. 5, pp. 9363–9373, 2012.

[NAO 13] NAOI K., NAOI W., AOYAGI S. *et al.*, "New generation 'nanohybrid supercapacitor'", *Accounts of Chemical Research*, vol. 46, no. 5, pp. 1075–1083, 2013.

[NÈG 15] NÈGRE L., DAFFOS B., TABERNA P.L. *et al.*, "Solvent-free electrolytes for electrical double layer capacitors", *Journal of the Electrochemical Society*, vol. 162, no. 5, pp. A5037–A5040, 2015.

[NIS 12] NISHIHARA H., KYOTANI T., "Templated nanocarbons for energy storage", *Advanced Materials*, vol. 24, pp. 4473–4498, 2012.

[NOH 06] NOHARA S., ASAHINA T., WADA H. *et al.* "Hybrid capacitor with activated carbon electrode, $Ni(OH)_2$ electrode and polymer hydrogel electrolyte", *Journal of Power Sources*, vol. 157, pp. 605–609, 2006.

[OBJ 17] OBJECTIF ECO, available at http://www.objectifeco.com/bourse/matieres-premieres/energies/bluetram-tramway-a-super condensateur-made-in-france.html, 2017.

[OHB 13] OHBA T., KANEKO K., "Competition of desolvation and stabilization of organic electrolytes in extremely narrow nanopores", *Journal of Physical Chemistry C*, vol. 117, pp. 17092–17098, 2013.

[OMA 12] OMAR N., DAOWD M., HEGAZY O. *et al.*, "Assessment of lithium-ion capacitor for using in battery electric vehicle and hybrid electric vehicle applications", *Electrochimica Acta*, vol. 86, pp. 305–315, 2012.

[PAN 06] PANDOLFO A.-G., HOLLENKAMP A.-F., "Carbon properties and their role in supercapacitors", *Journal of Power Sources*, vol. 157, pp. 11–27, 2006.

[PAN 12] PANDE P., RASMUSSEN P.-G., THOMPSON L.-T.-J., "Charge storage on nanostructured early transition metal nitrides and carbides", *Journal of Power Sources*, vol. 207, pp. 212–215, 2012.

[PAR 02] PARK J.-H., PARK O.-O., SHIN K.-H. *et al.*, "An electrochemical capacitor based on a $Ni(OH)_2$/activated carbon composite electrode", *Electrochemical and Solid-State Letters*, vol. 5, pp. H7–H10, 2002.

[PEA 14] PEAN C., MERLET C., ROTENBERG B. *et al.*, "On the dynamics of charging in nano-porous carbon-based supercapacitors", *ACS Nano*, vol. *8*, pp. 1576–1583, 2014.

[PEC 08] PECH D., GUAY D., BROUSSE T., *et al.*, "Concept for charge storage in electrochemical capacitors with functionalized carbon electrodes", *Electrochemical and Solid-State Letters*, vol. 11, no. 11, pp. A202–A205, 2008.

[PEL 04] PELL W.-G., CONWAY B.-E., "Peculiarities and requirements of asymmetric capacitor devices based on combination of capacitor and battery-type electrodes", *Journal of Power Sources*, vol. 136, pp. 334–345, 2004.

[PER 09] PERRET P., BROUSSE T., BÉLANGER D., *et al.*, "Electrochemical template synthesis of ordered lead dioxide nanowires", *Journal of the Electrochemical Society*, vol. 156, pp. A645–A651, 2009.

[PER 11] PERRET P., KHANI Z., BROUSSE T., *et al.*, "Carbon/PbO_2 asymmetric electrochemical capacitor based on methanesulfonic acid electrolyte", *Electrochimica Acta*, vol. 56, pp. 8122–8128, 2011.

[PET 00] PETERSSON I., AHLBERG E., "Oxidation of electrodeposited lead-tin alloys in 5 M H_2SO_4", *Journal of Power Sources*, vol. 91, pp. 143–149, 2000.

[PIE 14] PIERSON J.-F., MOREL A., CAPON F., "VN thin films as electrode materials for electrochemical capacitors", *Electrochimica Acta*, vol. 141, pp. 203–211, 2014.

[POG 11a] POGNON G., BROUSSE T., BÉLANGER D., "Effect of molecular grafting on the pore size distribution and the double layer capacitance of activated carbon for electrochemical double layer capacitors", *Carbon*, vol. 49, pp. 1340–1348, 2011.

[POG 11b] POGNON G., BROUSSE T., DEMARCONNAY L. *et al.*, "Performance and stability of electrochemical capacitor based on anthraquinone modified activated carbon", *Journal of Power Sources*, vol. 196, no. 8, pp. 4117–4122, 2011.

[POG 12] POGNON G., COUGNON C., MAYILUKILA D. *et al.*, "Catechol-modified activated carbon prepared by the diazonium chemistry for application as active electrode material in electrochemical capacitor", *ACS Applied Materials & Interfaces*, vol. 4, pp. 3788–3796, 2012.

[QUA 07] QUAN M., SANCHEZ D., WASYLKIW M.-F. *et al.*, "Voltammetry of quinones in unbuffered aqueous solution: reassessing the roles of proton transfer and hydrogen bonding in the aqueous electrochemistry of quinones", *Journal of the American Chemical Society*, vol. 129, pp. 12847–12856, 2007.

[RAT 14] RATAJCZAK P., JUREWICZ K., SKOWRON P. *et al.*, "Effect of accelerated ageing on the performance of high voltage carbon/carbon electro-chemical capacitors in salt aqueous electrolyte", *Electrochimica Acta*, vol. 130, pp. 344–350, 2014.

[RAY 06a] RAYMUNDO-PINERO E., KIERZEK K., MACHNIKOWSKI J. *et al.*, "Relationship between the nanoporous texture of activated carbons and their capacitance properties in different electrolytes", *Carbon*, vol. 44, pp. 2498–2507, 2006.

[RAY 06b] RAYMUNDO-PIÑERO E., LEROUX F., BÉGUIN F., "A high-performance carbon for supercapacitors obtained by carbonization of a seaweed biopolymer", *Advanced Materials*, vol. 18, pp. 1877–1882, 2006.

[RAY 13] RAYMUNDO-PIÑERO E., GAO Q., BÉGUIN F., "Carbons for supercapacitors obtained by one-step pressure induced oxidation at low temperature", *Carbon*, vol. 61, pp. 278–283, 2013.

[RAZ 98] RAZOUMOV S., KLEMENTOV A., LITVINENKO S. *et al.*, Asymmetric electrochemical capacitor and method of making, US patent 6,222,723, December 7, 1998.

[SAN 10] SANTOS-PENA J., CROSNIER O., BROUSSE T., "Nanosized alpha-$LiFeO_2$ as electrochemical supercapacitor electrode in neutral sulfate electrolytes", *Electrochimica Acta*, vol. 55, pp. 7511–7515, 2010.

[SAT 11] SATHIYA M., PRAKASH A.-S., RAMESHA K. *et al.*, "V2O5-anchored carbon nanotubes for enhanced electrochemical energy storage", *Journal of the American Chemical Society*, vol. 133, pp. 16291–16299, 2011.

[SCR 95] SCROSATI B., "Polymer electrodes", in BRUCE P. (ed.), *Solid State Electrochemistry*, Cambridge University Press, Cambridge, 1995.

[SHI 10] SHIM Y., KIM H.-J., "Nanoporous carbon supercapacitors in an ionic liquid: a computer simulation study", *ACS Nano*, vol. 4, pp. 2345–2355, 2010.

[SHU 09] SHUICHI I., YUICHIRO A., SHINYA M. *et al.*, "Degradation responses of activated-carbon-based EDLCs for higher voltage operation and their factors", *Journal of the Electrochemical Society*, vol. 156, pp. A563–A571, 2009.

[SHU 13] SHU D., LU C., CHENG F. *et al.*, "Enhanced capacitance and rate capability of nanocrystalline VN as electrode materials for supercapacitors", *International Journal of Electrochemical Science*, vol. 8, pp. 1209–1225, 2013.

[SIM 08a] SIMON P., GOGOTSI Y., "Materials for electrochemical capacitors: building a battery of the future", *Nature Materials*, vol. 7, pp. 845–854, 2008.

[SIM 08b] SIMON P., BURKE A., "Nanostructured carbons: double-layer capacitance and more", *Electrochemical Society Interface*, vol. 17, no. 1, pp. 38–43, 2008.

[SIM 13] SIMON P., GOGOTSI Y., "Capacitive energy storage in nanostructured carbon-electrolyte systems", *Account for Chemical Research*, vol. 5, pp. 1094–1103, 2013.

[SIM 14] SIMON P., GOGOTSI Y., DUNN B., "Where do batteries end and supercapacitors begin?", *Science Magazine*, vol. 343, pp. 1210–1211, 2014.

[SIV 10] SIVAKKUMAR S.-R., NERKAR J.-Y., PANDOLFO A.-G., "Rate capability of graphite materials as negative electrodes in lithium-ion capacitors", *Electrochimica Acta*, vol. 55, pp. 3330–3335, 2010.

[SMI 09] SMITH R.-D.-L., PICKUP P.-G., "Voltammetric quantification of the spontaneous chemical modification of carbon black by diazonium coupling", *Electrochimica Acta*, vol. 54, pp. 2305–2311, 2009.

[SNO 11] SNOOK G.-A., KAO P., BEST A.-S.-J., "Conducting-polymer-based supercapacitor devices and electrodes", *Journal of Power Sources*, vol. 196, no. 1, pp. 1–12, 2011.

[SON 07] SONG B., JIAN J.-K., WANG G., "Facile and general route to nitrides by a modified solid-state metathesis pathway", *Chemistry of Materials*, vol. 19, pp. 1497–1502, 2007.

[TAB 03] TABERNA P.-L., SIMON P., FAUVARQUE J.-F., "Electrochemical characteristics and impedance spectroscopy studies of carbon/carbon supercapacitor", *Journal of the Electrochemical Society*, vol. 150, no. 3, pp. A292–A300, 2003.

[TES 03] TESSIER F., MARCHAND R., "Ternary and higher order rare-earth nitride materials: synthesis and characterization of ionic-covalent oxynitride powders", *Journal of Solid State Chemistry*, vol. 171, pp. 143–151, 2003.

[TIA 10] TIAN Q., LIAN K., "Solid asymmetric electrochemical capacitors using proton-conducting polymer electrolytes", *Electrochemistry Communications*, vol. 12, pp. 517–519, 2010.

[TOU 04] TOUPIN M., BROUSSE T., BÉLANGER D., "Charge storage mechanism of MnO_2 electrode used in aqueous electrochemical capacitor", *Chemistry Materials*, vol. 16, no. 16, pp. 3184–3190, 2004.

[TOU 06] TOUPIN M., DELAHAYE T., BROUSSE T. *et al.*, "Nanostructured transition metal oxides for aqueous hybrid electrochemical supercapacitors", *Applied Physics A*, vol. 82, pp. 599–606, 2006.

[TOU 07] TOUPIN M., BÉLANGER D., "Thermal Stability study of aryl modified carbon black by *in situ* generated diazonium salt", *Journal of Physical Chemistry C*, vol. 111, no. 14, pp. 5394–5401, 2007.

[TSA 13a] TSAI W.-Y., GAO P.-C., DAFFOS B. *et al*, "Ordered mesoporous silicon carbide-derived carbon for high-power supercapacitors", *Electrochemistry Communications*, vol. 34, pp. 109–112, 2013.

[TSA 13b] TSAI W.-Y., LIN R., MURALI S. *et al.*, "Outstanding performance of activated graphene based supercapacitors in ionic liquid electrolyte from -50 to 80 °C", *Nano Energy*, vol. 2, pp. 403–411, 2013.

[VAR 95] VARAKIN I., STEPANOV A., MENUKHOV V., Capacitor with a double electrical layer, Patent WO97/07518 and US patent 5,986,876, August 14, 1995.

[VAR 98] VARAKIN I.-N., KLEMENTOV A.-D., LITVINENKO S.-V. *et al.*, "New ultracapacitors developed by JSC ESMA for various applications", *Proceedings of the 8th International Seminar on Double-Layer Capacitors and Similar Devices*, Deerfield Beach, December 1998.

[VIL 03] VILLERS D., JOBIN D., SOUCY C. *et al.* "The influence of the range of electroactivity and capacitance of conducting polymers on the performance of carbon/conducting polymer hybrid supercapacitor", *Journal of Electrochemical Society*, vol. 150, pp. A747–A752, 2003.

[VIX 04] VIX-GUTERL C., SAADALLAH S., JUREWICZ K. *et al.*, "Supercapacitor electrodes from new ordered porous carbon materials obtained by a templating procedure", *Materials Science and Engineering: B*, vol. 108, pp. 148–155, 2004.

[VOL 02] VOL'JKOVICH Y.-M., SERDYUK T.-M., "Electrochemical capacitors", *Russian Journal of Electrochemistry*, vol. 38, pp. 935–959, 2002.

[WAN 06] WANG Y.-G., XIA Y.-Y., "Hybrid aqueous energy storage cells using activated carbon and lithium-intercalated compounds. I. The C/$LiMn_2O_4$ system", *Journal of The Electrochemical Society*, vol. 153, no. 2, pp. A450–A454, 2006.

[WAN 08a] WANG D.-W., LI F., CHENG H.-M., "Hierarchical porous nickel oxide and carbon as electrode materials for asymmetric supercapacitor", *Journal of Power Sources*, vol. 185, pp. 1563–1568, 2008.

[WAN 08b] WANG H., YOSHIO M., "Performance of AC/graphite capacitors at high weight ratios of AC/graphite", *Journal of Power Sources*, vol. 177, pp. 681–684, 2008.

[WAN 10] WANG H., SANCHEZ CASALONGUE H., LIANG Y., "$Ni(OH)_2$ nanoplates grown on graphene as advanced electrochemical pseudocapacitor materials", *Journal of the American Chemical Society*, vol. 132, no. 21, pp. 7472–7477, 2010.

[WAN 11] WANG K., ZHAO P., ZHOU X.-M. et al., "Flexible supercapacitors based on cloth-supported of conducting array/SWCNT composites", *Journal of Material Chemistry,* vol. 21, no. 41, pp. 16373–16378, 2011.

[WEI 09] WEI W., CUI X., CHEN W. *et al.*, "Electrochemical cyclability mechanism for MnO_2 electrodes utilized as electrochemical supercapacitors", *Journal of Power Sources*, vol. 186, pp. 543–550, 2009.

[WEI 12a] WEI L., YUSHIN G., "Nanostructured activated carbons from natural precursors for electrical double layer capacitors", *Nano Energy*, vol. 1, pp. 552–565, 2012.

[WEI 12b] WEISSMANN M., CROSNIER O., BROUSSE T. *et al.*, "Electrochemical study of anthraquinone groups, grafted by the diazonium chemistry, in different aqueous media-relevance for the development of aqueous hybrid electrochemical capacitor", *Electrochimica Acta*, vol. 82, pp. 250–256, 2012.

[WEI 13] WEINGARTH D., FOELSKE-SCHMITZ A., KOETZ R., "Cycle *versus* voltage hold, which is the better stability test for electrochemical double layer capacitors?", *Journal of Power Sources*, vol. 225, pp. 84–88, 2013.

[WU 10] WU Q., XU Y., YAO Z. et al., "Supercapacitors based on flexible graphene/polyaniline nanofiber composite films", *ACS Nano*, vol. 4, no. 4, pp. 1963–1970, 2010.

[WU 14] WU W., YAN J.-Y., WISE A. *et al.*, "Using Intimate carbon to enhance the performance of $NaTi_2(PO_4)(3)$ anode materials: carbon nanotubes *versus* graphite", *Journal of the Electrochemical Society*, vol. 161, pp. A561–A567, 2014.

[XIA 13] XIAO X., PENG X., JIN H. *et al.*, "Freestanding mesoporous VN/CNT hybrid electrodes for flexible all-solid-state supercapacitors", *Advanced Materials*, vol. 25, no. 36, pp. 5091–5097, 2013.

[XIN 11] XING W., QIAO S., WU X. *et al.*, "Exaggerated capacitance using electro-chemically active nickel foam as current collector in electrochemical measurement", *Journal of Power Sources*, vol. 196, pp. 4123–4127, 2011.

[YAN 10] YAN J., WEI T., QIAO W. *et al.*, "Rapid microwave-assisted synthesis of graphene nanosheet/Co_3O_4 composite for supercapacitors", *Electrochimica Acta*, vol. 55, no. 23, pp. 6973–6978, 2010.

[YAN 13] YANG X., CHENG C., WANG Y. *et al.*, "Liquid-mediated dense integration of graphene materials for compact capacitive energy storage", *Science*, vol. 314, pp. 534-537, 2013.

[YU 09a] YU N., GAO L., "Electrodeposited PbO_2 thin film on Ti electrode for application in hybrid supercapacitor", *Electrochemistry Communications*, vol. 11, pp. 220–222, 2009.

[YU 09b] YU N., GAO L., ZHAO S. *et al.*, "Electrodeposited PbO_2 thin film as positive electrode in PbO_2/AC hybrid capacitor", *Electrochimica Acta*, vol. 54, pp. 3835–3841, 2009.

[ZHA 11] ZHANG L., HOLT C.-M.-B., LUBER E.-J. *et al.*, "High rate electrochemical capacitors from three-dimensional arrays of vanadium nitride functionalized carbon nanotubes", *Journal of Physical Chemistry C*, vol. 115, pp. 24381–24393, 2011.

[ZHE 03] ZHENG J.-P., "The limitations of energy density of battery and double-layer capacitor asymmetric cells", *Journal of The Electrochemical Society*, vol. 150, pp. A484–A492, 2003.

[ZHO 09a] ZHOU X., CHEN H., SHU D. *et al.*, "Study on the electrochemical behavior of vanadium nitride as a promising supercapacitor material", *Journal of Physics and Chemistry of Solids*, vol. 70, pp. 495–500, 2009.

[ZHO 09b] ZHOU W.-J., XU M.-W., ZHAO D.-D. *et al.*, "Electrodeposition and characterization of ordered mesoporous cobalt hydroxide films on different substrates for supercapacitors", *Microporous and Mesoporous Materials*, vol. 117, no. 1-2, pp. 55–60, 2009.

[ZHO 11] ZHOU X., SHANG C., GU L. *et al.*, "Mesoporous coaxial titanium nitride-vanadium nitride fibers of core-shell structures for high-performance supercapacitors", *ACS Applied Materials & Interfaces*, vol. 3, pp. 3058–3063, 2011.

[ZHU 11] ZHU Y., MURALI S., STOLLER M.-D. *et al.*, "Carbon-based supercapacitors produced by activation of graphene", *Science*, vol. 332, pp. 1537–1541, 2011.

[ZUK 05] ZUKALOVA M., KALBAC M., KAVAN L. *et al.*, "Pseudocapacitive lithium storage in TiO_2(B)", *Chemistry of Materials*, vol. 17, no. 5, pp. 1248–1255, 2005.

Index

A, B

C

D, E, F

G, H, I

L, M, N

O, P, Q

R, S

T, V

Other titles from

iSTE

in

Energy

2017

LA SCALA Massimo
From Smart Grids to Smart Cities: New Challenges in Optimizing Energy Grids

2016

ALLARD Bruno
Power Systems-on-Chip: Practical Aspects of Design

ANDRE Michel, SAMARAS Zissis
Energy and Environment

DUFOUR Anthony
Thermochemical Conversion of Biomass for the Production of Energy and Chemicals

2015

CROGUENNEC Laurence, MONCONDUIT Laure, DEDRYVÈRE Rémi
Electrodes for Li-ion Batteries
(Energy Storage – Batteries, Supercapacitors Set – Volume 2)

LEPRINCE-WANG Yamin
Piezoelectric ZnO Nanostructure for Energy Harvesting
(Nanotechnologies for Energy Recovery Set – Volume 1)

ROBYNS Benoît, FRANÇOIS Bruno, DELILLE Gauthier, SAUDEMONT Christophe
Energy Storage in Electric Power Grids

ROSSI Carole
Al-based Energetic Nanomaterials
(Nanotechnologies for Energy Recovery Set – Volume 2)

TARASCON Jean-Marie, SIMON Patrice
Electrochemical Energy Storage
(Energy Storage – Batteries, Supercapacitors Set – Volume 1)

2013

LALOUI Lyesse and DI DONNA Alice
Energy Geostructures: Innovation in Underground Engineering

2012

BECKERS Benoit
Solar Energy at Urban Scale

ROBYNS Benoît, DAVIGNY Arnaud, FRANÇOIS Bruno, HENNETON Antoine, SPROOTEN Jonathan
Electricity Production from Renewable Energies

2011

GAO Fei, BLUNIER Benjamin, MIRAOUI Abdellatif
Proton Exchange Membrane Fuel Cell Modeling

MULTON Bernard
Marine Renewable Energy Handbook

2010

BRUNET Yves
Energy Storage

2009

SABONNADIERE Jean-Claude
Low Emission Power Generation Technologies and Energy Management

SABONNADIERE Jean-Claude
Renewable Energy Technologies

Printed and bound by CPI Group (UK) Ltd, Croydon, CR0 4YY

13/09/2026

14961080-0002